Surface Characteristics
of Fibers and Textiles

(in two parts)

Part II

FIBER SCIENCE SERIES

Series Editor

L. REBENFELD

Textile Research Institute
Princeton, New Jersey

Other Volumes in Preparation

Surface Characteristics of Fibers and Textiles

(in two parts)

Part II

edited by

M. J. Schick

Diamond Shamrock Corporation
Process Chemicals Division
Morristown, New Jersey

CRC Press
Taylor & Francis Group
Boca Raton London New York

CRC Press is an imprint of the
Taylor & Francis Group, an **informa** business

First published 1977 by Marcel Dekker, Inc

Published 2019 by CRC Press
Taylor & Francis Group
6000 Broken Sound Parkway NW, Suite 300
Boca Raton, FL 33487-2742

© 1977 by Taylor & Francis Group, LLC
CRC Press is an imprint of Taylor & Francis Group, an Informa business

First issued in paperback 2019

No claim to original U.S. Government works

ISBN-13: 978-0-367-45208-7 (pbk)
ISBN-13: 978-0-8247-6531-6 (hbk)

Visit the Taylor & Francis Web site at
http://www.taylorandfrancis.com

and the CRC Press Web site at
http://www.crcpress.com

Library of Congress Cataloging in Publication Data (Revised)

Main entry under title:

Surface characteristics of fibers and textiles.

    Includes bibliographical references.
    1. Textile fibers. 2. Textile fabrics.
I. Schick, Martin J.
TS1449.S79   677'.0283   75-10346

ISBN 0-8247-6531-1

CONTENTS

CONTENTS

LIST OF CONTRIBUTORS

G. GRAHAM ALLAN, Department of Chemical Engineering, College of Forest Resources, University of Washington, Seattle, Washington

T. H. GRINDSTAFF, Textile Fibers Department, Fiber Surface Research Section, E. I. du Pont de Nemours & Company, Inc., Kinston, North Carolina

PERCY GROSBERG, Department of Textile Industries, University of Leeds, Leeds, England

J. E. LAINE, Department of Chemical Engineering, College of Forest Resources, University of Washington, Seattle, Washington

BERNARD M. LICHSTEIN, Patient Care Division, Johnson & Johnson, New Brunswick, New Jersey

BERNARD MILLER, Textile Research Institute, Princeton, New Jersey

A. N. NEOGI, Department of Chemical Engineering, College of Forest Resources, University of Washington, Seattle, Washington

H. T. PATTERSON, Textile Fibers Department, Fiber Surface Research Section, E. I. du Pont de Nemours & Company, Inc., Kinston, North Carolina

MALCOLM E. SCHRADER, Materials Department, David W. Taylor Naval Ship Research and Development Center, Annapolis, Maryland

CONTENTS OF PART I

PREFACE

Within the past two decades the phenomenal growth in the production of man-made fibers has focused interest on the surface properties of fibers and textiles. A perusal of the literature indicates that papers dealing with the surface properties of fibers were relatively sparse until the 1950's. However, more sophisticated probings into the fundamental concepts are now in progress. Coupled with this research program has been a continuing effort to apply the results of this research to practical problems. Although some of these developments have been covered in other books, the need for a comprehensive treatise on the surface characteristics of fibers and textiles has been felt for a long time. Our treatise endeavors to fill this need by presenting the substantial progress in science and technology related to the surface properties of fibers and textiles. The investigation of these surface characteristics requires an interdisciplinary approach which combined surface science with textile technology and eliminates the myopia of classical-discipline centered work so commonly seen in the past.

Since the subjects to be discussed represent such a broad spectrum, no single person could write a critical review on more than a very limited number of topics. Thus, the treatise is a collection of chapters written by outstanding specialists. The contributions, although self-contained, are interrelated. This treatise offers critical reviews, which describe experimental facts, theories, and processes and which handle these in a rigorous way. Symbols are clearly defined in each chapter. It was impossible to publish a work of this magnitude with all chapters in a logical order and in a single book. However, every effort was made to group related chapters in sequence. A certain amount of overlap is unavoidable, but was kept to a minimum. The indexes to the entire book appear at the end of Part II.

The treatise encompasses the surface properties of both natural and synthetic fibers. Emphasis has been placed on the frictional, geometrical, electrical, wetting, adhesive, and optical properties of fibers and fabrics as well as on phenomena related to these properties. In addition, experimental procedures used to assess the surface properties of fibers and fabrics are described. It is the sincere hope of the contributors and editor that much exciting and stimulating research will result from this effort. Likewise, it is hoped that this presentation of the most up-to-date thinking in this area of surface characteristics of fibers and textiles will lead to further advances in textile technology.

It is with great pleasure that I express my sincere gratitude to the contributing authors in carefully preparing their chapters and their splendid cooperation with the editor. Likewise, the guidance and enthusiastic support of Dr. Ludwig Rebenfeld, consulting editor of the Fiber Science Series, is gratefully acknowledged. Thanks are also due to the academic, government, and industrial organizations with which the contributing authors and the editor are associated for the assistance given to us, and to the publishers who have given their permissions for the reproduction of illustrations. Finally, thanks are expressed to the following staff members of the Diamond Shamrock Corporation: Miss Marion Kearney and Mrs. Evelyn Fornale for secretarial assistance and Miss Sandy Haines for preparation of illustrations.

M. J. Schick
Morristown, New Jersey

Surface Characteristics of Fibers and Textiles

(in two parts)

Part II

Chapter 11

THE WETTING OF FIBERS

Bernard Miller

Textile Research Institute
Princeton, New Jersey

I. INTRODUCTION

Fiber scientists and textile technologists have a pervading need to understand and measure the wetting of fibers and filaments. Wetting of fibrous materials can critically affect many manufacturing processes, as well as the end-use performance of materials. In addition, certain manifestations of wetting behavior can be used to monitor, at least on a relative basis, changes in the surface free energy of a material. Indeed, the measurement of spontaneous surface wetting is one of the few experimental tools available for the study of solid-surface energetics.

For fibers as well as for other solids, wetting phenomena can be divided into two general classes: equilibrium wetting, where liquid and solid phases once placed in contact are no longer externally perturbed; and dynamic wetting, where the liquid or solid (or both) is kept in motion relative to the other phase throughout the wetting process.

As might be expected, no single theory or experimental technique can deal with both of these cases, and they will be discussed separately in this chapter. As an additional restriction, this writing will deal only with the wetting of single-fiber elements and not fiber assemblies.

II. THE MEANING OF WETTING AND WETTABILITY

If one had the opportunity to carry out a word association test on a typical cross section of technically trained individuals, the response to the term "wettability" (or "wetting") would most likely be "contact angle." Indeed, the habit of describing the physical interaction of liquids and solids in terms of contact angles has become so widespread that a reappraisal of its limitations is urgently needed. Many recent papers dealing with wetting phenomena suggest that it is not generally understood that the numerical value of a contact angle or its cosine does not bear a simple relationship to the surface free energy of a solid or the energetics of interaction between a solid and a liquid. Brewis [1] has pointed this out in his discussion of the significance of the observed relative insensitivity of contact angle to temperature, such as has been reported by Johnson and Dettre [2]. Therefore, this review includes an attempt to define and illustrate what the contact angle does and does not mean, and what other quantities should be determined for the proper analysis and understanding of wetting phenomena.

In addition, it will be helpful to maintain a clear distinction between two terms that are sometimes used interchangeably, namely "wetting" and "wettability." The wetting of a solid surface is understood to be the condition resulting from its contact with a specified liquid under specified conditions. Wettability is the potential of a surface to interact with liquids with specified characteristics.

III. EQUILIBRIUM WETTING

A. General Considerations

The equilibrium condition for the wetting of a solid surface by a liquid is actually a three-phase equilibrium of solid, liquid, and vapor. Looked at in two dimensions, as in Figure 1, one can describe the equilibrium point of contact as the intersection of three interfaces: solid-liquid, liquid-vapor, and solid-vapor. In three dimensions this intersection becomes a line of contact (Fig. 2). For many solids this line of contact is straight, but with fibers it is obviously highly curved. However, this curvature has no effect on the inherent wettability of a fiber surface, as will be demonstrated later.

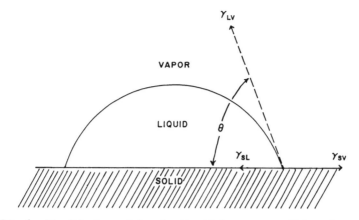

FIG. 1.   Equilibrium state of a liquid drop on a solid surface.

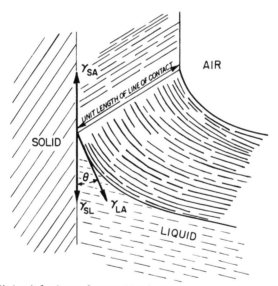

FIG. 2.   Pictorial view of a vertical surface in contact with a liquid.

It is convenient to make use of the concept of surface tension (γ, usually in dynes/cm) to describe the balance of forces on this three-phase intersection. Since surface tension can be defined as the attractive force of one unit for another in the plane of an interface, the solid-vapor interface (referring to Figure 1) will pull the line of contact to the right with a force γ_{SV}; the solid-liquid interface will pull to the left with a force γ_{SL}; and the liquid-vapor interface will pull to the left with a force $\gamma_{LV} \cos \theta$. Thus the equilibrium condition is described by the Young-Dupré [3] equation, developed in the nineteenth century:

$$\gamma_{SV} - \gamma_{SL} = \gamma_{LV} \cos \theta \tag{1}$$

The angle θ is called the contact angle and serves as a convenient means for visualizing and describing the geometry of solid-liquid contact. However, Adam [4] reminds us that the equilibrium condition cannot exist when the contact angle is zero, and the above equation does not apply. Some writers have defined wetting as synonymous with a zero contact angle and nonwetting as any condition with a contact angle greater than 90° [5]. This is unfortunate since it implies that wetting occurs only in nonequilibrium situations and leaves the important region between 0° and 90° in limbo. As pointed out by Huntsberger [6], the distinction between zero and nonzero contact angles is really the distinction between spreading and nonspreading of a liquid across a surface. Whether or not it spreads, there is always some wetting when a liquid comes in contact with a surface (with one possible exception, which will be discussed later).

Most likely because of its convenience, many workers have fallen into the habit of using the contact angle or its cosine function as a direct measure of wettability or degree of wetting. For comparisons between systems, especially when different liquids are involved, this can be extremely misleading. The contact angle is not the cause of wetting but its consequence, and is determined by the net effect of three essentially independent attractive forces [see Eq. (1)]. Thus an observed change in $\cos \theta$ cannot always be simply interpreted. If two solids are compared using the same wetting liquid, a difference in contact angle will tell which solid has the greater attraction for the liquid. If two liquids are compared with respect to the same solid, any difference in contact angle has no clearcut significance by itself, unless adjusted in some manner for the corresponding liquid-liquid interactions. In either case, changes in $\cos \theta$ usually can indicate only relative differences.

To get the most information from available data and a better understanding of wetting phenomena the surface scientist should not be concerned with the contact angle per se, but only as it can serve to obtain knowledge of more fundamental and useful quantities. One of these is γ_{SV}, the surface energy of the solid in question, a property which should reflect the chemical nature and physical structure of the solid surface. Another is the net attraction between solid and liquid, usually called the work of adhesion, W_{SL}. This too is dependent in part on the nature of the solid surface. In many instances where a property or process is influenced by wetting behavior, it is the work of adhesion that is the critical factor. The recent work of Kaelble and his co-workers [7-9] represents a good example of this approach.

B. The Surface Free Energy of a Solid

Since there is no direct method available for measuring γ_{SV}, a good
deal of effort has been concentrated on indirect ways to obtain it or some
quantity proportional to it. The method in most general use is that introduced by Zisman and co-workers [10-12], in which contact-angle
measurements are made on the solid of interest with a homologous series
of liquids. A linear relationship between cos θ and the surface tensions
of the liquids is assumed and this relationship is extrapolated to $\theta = 0$
(cos $\theta = 1$). The corresponding limiting value of the surface tension is
defined as the "critical surface tension" of the solid surface γ_c. The
rationale for this appears to be that maximum wetting ($\theta = 0$) should occur when liquid and solid have the same surface tensions, but it is not
apparent why this should be so. Many workers have expressed doubts
about the use of this (essentially) empirical approach to obtain a fundamental thermodynamic property. One obvious objection to it is the demonstrated fact that γ_c is not independent of the polarities of the liquid
phase [13]. In addition, Neumann [14] has described in detail how the
relationship between cos θ and γ_{LV} can become extremely curvilinear
as θ approaches zero. A careful study of the antecedents of the Zisman
method for obtaining critical surface tensions shows that the original de-
cision to use a plot of cos θ versus γ_{LV} was quite arbitrary and was
evaluated with data covering only a very small range of surface tensions
[10]. No physical or mathematical arguments were offered for this
method of obtaining γ_c. The continued use of this method, both by its
originators and others, can only be considered unfortunate.

In contrast, Fowkes [15] has presented an analysis showing that
cos θ should be proportional to $1/\gamma_{LV}$, and there is considerable theoretical and experimental evidence to show that this is probably correct.
The following general argument supports this point of view: Consider
that

γ_{SV} is a fixed property of a solid surface.

γ_{LV} is a fixed property of a liquid surface.

W_{SL} and γ_{SL} are properties dependent on solid-liquid interaction,
but are themselves interdependent.

If we imagine a solid-liquid interface at which the solid does not
perturb the liquid (i.e., there is no solid-liquid attraction), it would be
true that $\gamma_{LV} \cong \gamma_{SL}$, since the liquid surface adjacent to the solid would
be no different from the liquid-air one. On the other hand, γ_{SL} will be

less than γ_{LV} if there is any pull of solid on liquid. As a first approximation let us assume that in such a case

$$\gamma_{SL} = \gamma_{LV} - kW_{SL} \qquad \text{(k constant)} \tag{2}$$

The general relationship between the surface tensions and work of adhesion as developed by Dupré [4] is

$$W_{SL} = \gamma_{LV} + \gamma_{SV} - \gamma_{SL} \tag{3}$$

Combining Equations (2) and (3),

$$W_{SL} = \gamma_{LV} + \gamma_{SV} - \gamma_{LV} + kW_{SL}$$

or

$$W_{SL}(1 - k) = \gamma_{SV} \tag{4}$$

Work of adhesion is related to cos θ [see Eq. (8)], so that

$$\gamma_{LV}(1 + \cos \theta)(1 - k) = \gamma_{SV}$$

and

$$\cos \theta = \frac{\gamma_{SV}}{1 - k} \frac{1}{\gamma_{LV}} - 1 \tag{5}$$

Therefore, a plot of cos θ against the reciprocal of liquid surface tension should be a straight line with a positive slope if $k < 1$. The plot will have its lowest point (cos θ = -1) when $1/\gamma_{LV}$ = 0, and a critical surface tension can be obtained by extrapolating to cos θ = 1. This result is very much like that developed by Fowkes [15], but without any restriction on the types of attractive forces involved.

Over and above any method of obtaining γ_c is the question of what it signifies. Fowkes [15] showed that γ_c could be equal to γ_{SV} only when the attractive forces involved were solely dispersive (e.g., no polar or hydrogen bonding). Only a relatively minor number of solid-liquid combinations meet this criterion, and it might be best if the concept of critical surface tension were retired in toto.

Other investigators have suggested ways to obtain a value for γ_{SV} that would be independent of the nature of any liquid phase. One that seems reasonable was introduced by Girifalco and Good [16], who developed an interaction parameter ϕ, defined as

$$\phi = \frac{\gamma_{SV} + \gamma_{LV} - \gamma_{SL}}{2 \sqrt{\gamma_{SV}\gamma_{LV}}} \tag{6}$$

This, when combined with Equation (1), gives two relationships containing the unknowns γ_{SV} and γ_{SL}. If ϕ can be determined, it would be possible to eliminate γ_{SL} and determine a value for γ_{SV}. Good and others used known molecular properties of materials to calculate values of ϕ. Neumann [14] has discussed several methods whereby experimental data can be used to obtain γ_{SV}. A similar approach using wetting data obtained with two liquids whose dispersive and nondispersive contributions to their surface tensions are known (i.e., water and methylene iodide) has been utilized by El-Shimi and Goddard [17], using the theory of Wu [18] to define γ_{SL}. This seems to be the most promising approach at this time for obtaining solid-surface free energies.

C. Work of Adhesion Between Solid and Liquid

The basic energetic parameter describing the interaction of a liquid and solid—the one that is capable of clearly quantifying what we call degree of wetting—is the work of adhesion W_{SL}. This is the total attraction per unit area between the two phases and is the result of both chemical and physical factors. Work of adhesion data can give a much clearer picture of the differences between systems being evaluated for wetting characteristics.

The work of adhesion between a solid and a liquid can be defined in general thermodynamic terms [i.e., Equation (3)]:

$$W_{SL} = \gamma_{LV} + \gamma_{SV} - \gamma_{SL}$$

Similarly, the corresponding binding of a liquid to itself is measured by the work of cohesion (W_{LL}) equal to twice the attraction of its units:

$$W_{LL} = 2\gamma_{LV} \tag{7}$$

Combining Equations (1) and (3) results in an expression for work of adhesion in terms of two measurable quantities, the surface tension of the liquid and the contact angle:

$$W_{SL} = \gamma_{LV} + \gamma_{LV} \cos \theta \tag{8}$$

Thus, the work of adhesion is equivalent to the sum of two quantities, one of which is a property of the liquid alone, so that the other must reflect the interaction between liquid and solid. Some investigators have preferred to use this latter term, $\gamma_{LV} \cos \theta$, as a measure of the degree of wetting, calling it adhesion tension [19] or specific wettability [20]. As will be shown, this quantity is quite useful but cannot serve as a complete wetting parameter.

For wetting with a given liquid, $\gamma_{LV} \cos \theta$ will change systematically with changes in solid-surface characteristics. However, the critical factor controlling the orientation of a solid-liquid interface is the difference between W_{SL} and W_{LL}, sometimes called the spreading coefficient [21]. How this difference determines the resultant effect of liquid-solid contact is shown in the following analysis:

For Spreading: Equilibrium spreading requires that solid-liquid attraction must equal or exceed the liquid-liquid attraction, that is,

$$W_{SL} - W_{LL} = 0 \tag{9}$$

Replacing with equivalent terms,

$$\gamma_{LV} + \gamma_{LV} \cos \theta - 2\gamma_{LV} = \gamma_{LV} \cos \theta - \gamma_{LV} = 0 \tag{10}$$

or

$$\gamma_{LV} \cos \theta = \gamma_{LV} \tag{11}$$

Therefore, spreading corresponds to $\cos \theta = 1$.

For Wetting: Wetting occurs whenever there is contact between liquid and solid. It is convenient to divide wetting into three subcategories which are best illustrated for the case where a liquid encounters a vertical solid surface:

(1) For positive wetting (the liquid rises up the solid surface),

$$0 < \cos \theta < 1$$

$$0 < \gamma_{LV} \cos \theta \leqq \gamma_{LV}$$

and since

$$W_{SL} - W_{LL} = \gamma_{LV} \cos \theta - \gamma_{LV}$$

$$-\gamma_{LV} < (W_{SL} - W_{LL}) < 0 \tag{12}$$

That is, positive wetting occurs when the work of cohesion exceeds the work of adhesion by less than $\frac{1}{2} W_{LL}$ (i. e., γ_{LV}).

(2) For zero displacement wetting (the liquid surface is not distorted upon contact with the solid),

$$\cos \theta = 0$$

$$\gamma_{LV} \cos \theta = 0$$

$$W_{SL} - W_{LL} = \gamma_{LV} \cos \theta - \gamma_{LV} = -\gamma_{LV} \tag{13}$$

and

$$W_{SL} = 2\gamma_{LV} - \gamma_{LV} = \gamma_{LV} \tag{14}$$

For this singular case the solid-liquid attraction is one half the liquid-liquid attraction. In descriptive terms, one can imagine that a liquid unit at the interface is attracted with equal force to the adjacent solid and to other liquid units.

(3) For negative wetting (liquid level is lowered at the solid-liquid interface)

$$-1 \leqq \cos \theta < 0$$

$$-\gamma_{LV} \leqq \gamma_{LV} \cos \theta < 0$$

and

$$-2\gamma_{LV} \leqq (W_{SL} - W_{LL}) < -\gamma_{LV} \tag{15}$$

Or, the attraction of the solid for the liquid is less than that for the liquid to itself up to a maximum difference equal to $2\gamma_{LV}$. Actually, there is no apparent reason why the work of cohesion could not exceed the work of adhesion by more than $2\gamma_{LV}$, which would truly be a case of non-wetting. However, such a condition has not been reported by any investigator.

A number of conclusions can be drawn from the preceding analysis:

1. Positive, zero-displacement, and negative wetting are all aspects of the same phenomenon, the result of competitive attractions of the solid surface and the liquid for liquid surface units.

2. Which wetting orientation prevails will depend on the relative magnitudes of the two attractions, not their absolute values.

3. A change in liquid will produce changes in both attractions, but not necessarily to the same degree.

4. A change in the solid surface will alter W_{SL} but not W_{LL}.

In summary, the two quantities which should be the object of any study of liquid-solid equilibrium interactions are the work of adhesion W_{SL}, and the solid-surface free energy γ_{SV}. Determining either of these quantities can depend on the accuracy and precision of experimental measurements of liquid surface tension γ_{LV}, and contact angle θ. Alternatively, the product of these two quantities (i. e., $\gamma_{LV}\cos\theta$), may be obtained directly, eliminating the need for a contact-angle measurement, as will be shown.

D. Experimental Considerations

1. Direct Contact-Angle Measurement

Most of the established experimental techniques for evaluating surface wetting properties and free energies have been developed for use with flat, isotropic surfaces. Under such conditions the observation and measurement of a contact angle when a drop of liquid is placed on a horizontal solid surface is not difficult to accomplish. As long as it makes no difference which profile direction is observed (that is, when the surface free energy of the solid is the same in all directions), direct or photographic measurements can be made with reasonable precision. A recent paper by Bluhm [22] describes how a highly magnifying optical microscope can be adapted for such purposes. However, even for such simple substrates, measurements of very small or very large angles are generally conceded to be of limited reliability. Rather than attempt a direct observation, many investigators have preferred to use techniques such as the tilted plate method [5], provided that the material to be studied is in suitable form.

Making contact angle measurements on filamentous materials is a much more difficult experimental problem. The direct approach requires that a drop of liquid be placed on a horizontally mounted sample and the contact angle observed from a point in the same horizontal plane and perpendicular to the long axis of the sample. However, this type of experiment has several pitfalls. Schwartz and his co-workers [23, 24] demonstrated how certain liquids have the tendency to completely surround a single fiber and form a symmetrical unduloid shape, while others can remain on one side of a fiber with a "clamshell" profile. In some instances, the same liquid could be made to adopt either of these configurations depending on how it was deposited. Yamaki and Katayama have described the significant effects of drop size and filament diameter on contact angles [25]. Such observations have deterred most prudent investigators from attempting direct contact-angle measurements. Instead, the preferred technique has been a version of the tilted-plate concept, using a small reservoir of liquid which is pierced by the fiber and an arrangement whereby it can be tilted relative to the liquid surface [26, 27]. Considerable precaution must be taken to make sure that the true contact angle can be observed and measured [28].

Even with surfaces which are conducive to precise contact angle measurements it is important to keep in mind the limitations inherent in such a method when one's goal is the determination of such properties as work of adhesion, solid-surface free energy, or even critical surface tension. Let us consider how the typical precision of "good" contact-angle measurements will influence the accuracy of the cos θ value, whose magnitude is involved in obtaining each of the above properties. Figure 3, a plot of cos θ as a function of θ, demonstrates the percentage uncertainty in the former resulting from a ± 1° error in measuring θ between 0° and 90°. (Such precision for a contact-angle measurement would be considered quite satisfactory by experienced workers.) For angles up to about 45° the uncertainty of the cosine value is not a serious problem, but with larger angles it is much more difficult to obtain an accurate value for cos θ. This includes angles up to 135°, since the cosine function is symmetrical about 90°.

If the objective of an experimental investigation is information on work of adhesion, the question arises as to whether the effort to obtain accurate contact angles is worth the trouble. Figure 4 is a plot of the relationship between work of adhesion and contact angle as mandated by Equation (8). Several points are obvious:

1. The same contact angle can represent quite different W_{SL} values.

2. If the wetting liquid has a very low surface tension, the precise measurement of θ is not critical.

3. If θ is small, perturbations of the liquid surface tension (or error in its measurement) will have a large effect on W_{SL}.

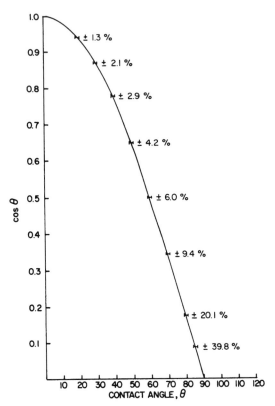

FIG. 3. Cosine θ as a function of θ, demonstrating the uncertainty in the former corresponding to a ±1° uncertainty in θ. (Reprinted from Ref. 20, p. 359, by courtesy of Textile Research Institute.)

Equation (8) can also be plotted with the liquid surface tension as independent variable (Fig. 5). This leads to additional useful observations:

4. W_{SL} is practically independent of contact angle between 0° and 30°.

5. There is no special significance to a 90° contact angle.

2. Wetting Force Measurements

An alternate experimental approach which does not require the direct measurement of contact angle appears to be promising, especially for dealing with fibers. The technique, based on the Wilhelmy balance principle [29], was first described by Collins [30] and its mathematical validity demonstrated by Allan [31]. Its use and implications have been discussed by other authors [2, 8, 32]. Bendure [33] has applied

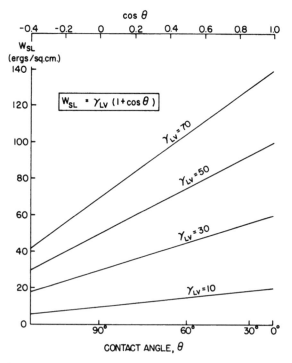

FIG. 4. Work of adhesion for several values of γ_{LV} as a function of the contact angle, according to the Young-Dupré equation.

it to the study of the dynamic wetting behavior of fibers at various oil and water interfaces. A first report of the version of this technique used at Textile Research Institute was published in 1975 [20].

According to the formula of Wilhelmy [29], the pull exerted on a vertical rod (i. e., filament or fiber) inserted into a liquid is expressed by

$$F_w = P\gamma_{LV} \cos \theta \tag{16}$$

where P is the perimeter of the solid along the three-phase boundary line and the other terms have their usual meaning. This pull can be downward (when the liquid rises up the solid) or it can be a "push" upward (when the meniscus is depressed at the interphase boundary). Normalizing for the length of contact between filament and liquid, we can describe a characteristic specific wetting force,

$$w = \frac{\text{force per filament}}{\text{perimeter of filament}} = \frac{F_w}{P} = \gamma_{LV} \cos \theta \tag{17}$$

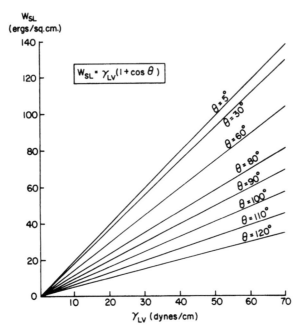

FIG. 5. Work of adhesion for several values of θ as a function of liquid surface tension.

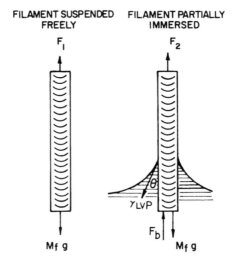

FIG. 6. Vertical forces on a suspended filament. If the buoyancy force F_b is negligible, $F_2 - F_1 = \gamma_{LV} P \cos \theta$ ($M_f g$ = gravitational force).

Thus, the determination of w, which has been called adhesion tension [19] or specific wettability [20], will give a value for $\gamma_{LV} \cos \theta$ without any actual measurement of θ. Combined with an independently obtained value for γ_{LV}, this can be used to calculate work of adhesion [using Eq. (8)] or $\cos \theta$. The determination of the wetting force F_w can be carried out by arranging to measure the change in weight of a vertical fiber (e.g., with a microbalance) that occurs when it is placed in contact with a liquid.

Figure 6 illustrates this concept in terms of the forces acting on the fiber before and after contact. In addition to the wetting force, any significant immersion of the fiber in the liquid will induce a buoyancy force F_b. Conceivably, this could be avoided by not allowing the fiber to penetrate below the liquid surface, but this is highly inadvisable. For one thing, the measurement then will be inordinately influenced by the shape of the fiber end. For another, it is extremely difficult to reproduce this type of contact. Thirdly, this limits the measurement to one location on the fiber surface, one that is very likely not representative of the rest of the sample.

A much better technique is to make use of the buoyancy effect rather than trying to avoid it and, when necessary, to eliminate the immersion of fractured fiber ends by mounting the sample in the form of a rectangular loop with its ends placed where they will not come into contact with the liquid (see Fig. 7).

The balance of forces on a partially immersed filament is

$$F = F_w - F_b \tag{18}$$

where

F = balance reading corrected for reading in air (in dynes)

F_w = weight of liquid clinging to the filaments (i.e., the Wilhelmy wetting force)

F_b = buoyancy force

The buoyancy force can be calculated from the volume of solid immersed and the density of the liquid:

$$F_b = LANd + \ell Ad + B_s \tag{19}$$

where

L = depth of immersion of vertical filaments

A = cross-sectional area per filament

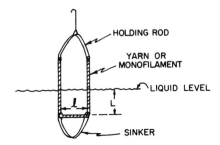

FIG. 7. Sample holder for nonrigid filaments and yarns. (Reprinted from Ref. 20, p. 360, by courtesy of Textile Research Institute.)

N = number of vertical filaments

d = density of liquid

ℓ = horizontal immersed filament length (see Fig. 7)

B_s = buoyancy force of the combined submerged holder and sinker

The specific wetting force w, as defined by Equation (17), becomes

$$w = \frac{F_w}{NP} = \frac{F + F_b}{NP} \tag{20}$$

Substituting for F_b from Equation (19),

$$w = \frac{1}{NP} (F + LANd + \ell Ad + B_s) \tag{21}$$

and, rearranging,

$$F = (NPw - \ell Ad - B_s) - LANd \tag{22}$$

or

$$F = I + SL \tag{23}$$

where the constant terms are

$$I = NPw - \ell Ad - B_s$$

$$S = -ANd$$

Thus, for a linear plot of observed force F as a function of depth of immersion L, the intercept is equal to I, and the slope is equal to S. The specific wetting force is then

$$w = \frac{1}{NP} (I + \ell Ad + B_s) \tag{24}$$

and the cross-sectional area of the filament or yarn is

$$A = \frac{-S}{Nd} \tag{25}$$

Combining Equations (24) and (25), we get

$$w = \frac{1}{NP} (I - \frac{\ell S}{N} + B_s) \tag{26}$$

Therefore, w can be obtained from the slope and intercept of a plot of F versus L without any direct measurement of the cross-sectional area of the sample. The buoyancy of the wire sinker can be determined experimentally or calculated from its dimensions and the density of the liquid. The horizontal distance ℓ is easily measured and will be another constant set by the holder dimensions. For round fiber cross sections, the perimeter P is obtained from diameter measurements; cross-sectional micrographs can be utilized to obtain perimeters for noncylindrical samples.

A single, sufficiently rigid filament can be used without any lower support or sinker, if liquid diffusion through its fractured end does not interfere with the wetting force measurement. In most cases this will be possible since the surface wetting measurement can be made quickly, long before any longitudinal diffusion occurs. For this case, Equation (22) can be simplified to

$$F = Pw - LAd \tag{27}$$

or, once again,

$$F = I + SL$$

where

$$I = Pw$$

and

$$S = -Ad$$

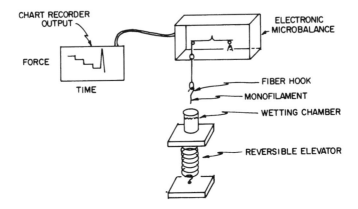

FIG. 8. Apparatus for wetting-force measurements. (Reprinted from Ref. 20, p. 361, by courtesy of Textile Research Institute.)

Now the specific wetting force on the filament will be

$$w = \frac{I}{P} \tag{28}$$

and its wetted cross section

$$A = -\frac{S}{d} \tag{29}$$

Thus, w, which is equivalent to $\gamma_{LV} \cos \theta$, can be obtained without measuring surface tension, contact angle, or the cross-sectional area of the wet sample. The latter can actually be determined from the experimental data, as was originally pointed out by Collins [30].

3. Experimental Details

An example of an apparatus which can be used for wetting-force measurements is shown in Figure 8. The sample is suspended from the microbalance hangdown wire, and its weight in air zeroed out before contact. The liquid is slowly raised until a force change is observed as a result of first contact with the suspended sample. This force change is then used only to mark the level of zero immersion. The liquid is then raised further a measured distance to produce a specific depth of sample immersion. The change in weight is recorded just long enough to insure that an equilibrium reading has been obtained (a few seconds is usually sufficient). The measurement is repeated at increasing depths of immersion until at least a

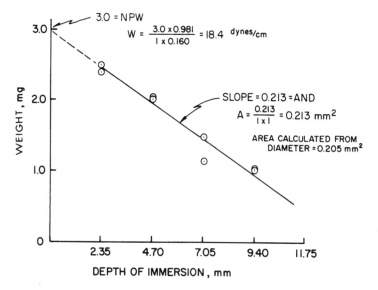

FIG. 9. Plot of wetting-force data for a polyester monofilament (0. 51 mm diameter). (Reprinted from Ref. 20, p. 362, by courtesy of Textile Research Institute.)

centimeter of filament has been traversed. The direction of liquid movement can then be reversed to obtain data for receding wetting in a similar manner (this will be discussed in detail later).

The observed force (i. e. , weight) changes are plotted as a function of depth of immersion, as shown in Figure 9. If the wettability and perimeter of the sample are constant along its length, then all the data will form a straight line which can be extrapolated to zero depth of immersion, thus correcting for buoyancy. If a single reading falls significantly far off from the straight-line relationship, it is undoubtedly due to a trivial surface defect or impurity at a singular spot and should be disregarded. This is one of the major advantages of this technique, since it eliminates the possibility of a single anomalous result being taken as representative. If most of the experimental points fall outside the statistically probable range of a liner plot or if no such relationship can be observed, then the sample does not have a uniform surface and any wettability measurement will be valueless.

As shown in Figure 9, the extrapolated weight (converted to dynes by multiplying by 0. 981 dyne/mg) is used to calculate the specific wettability w. The slope of the plot can be used to determine the cross-sectional area of the wetted filament. For smooth, round filaments this area usually compares quite well with that of the dry filament; however in some instances the wet value is greater. This latter result

TABLE 1

Work of Adhesion Compared with Contact Angle[a]

		EtOH (95%)	Toluene	Ethylene Glycol	Water
Nylon	W_{SL}[b]. . .	43.6	43.9	73.5	95.7
	(θ)	(18°)	(57°)	(57°)	(71°)
Polyester	W_{SL}. . . .	42.6	44.3	70.4	90.9
	(θ)	(26°)	(56°)	(61°)	(75°)
Polypropylene	W_{SL}. . . .	37.7	–	61.0	77.4
	(θ)	(47°)	–	(74°)	(86°)

[a]Both obtained from the same wetting-force data.
[b]Erg/cm^2.

is most likely a reflection of unwetted surface crevices which would increase the apparent buoyancy of the sample.

By reversing the experiment, data can be obtained for the equilibrium between liquid and pre-wet fiber surface. The level of liquid is lowered from its highest point on the sample and a series of force readings for different depths of immersion made as before. In most instances the observed force for a specific depth will be appreciably greater for the receding mode than for the advancing mode. This approach is by far the most reliable way to obtain information on this hysteresis effect.

4. Examples of Experimental Results

Table 1 contains wetting data for three different filaments tested with four liquids. Experimentally determined values for $w = \gamma_{LV}$ cos θ and γ_{LV} were used to obtain work of adhesion values ($W_{SL} = \gamma_{LV} + \gamma_{LV}$ cos θ) and contact angles. This tabulation illustrates dramatically how misleading contact-angle comparisons can be. Toluene and ethylene glycol form the same contact angle on nylon, yet the attraction of the latter liquid to this polymer is actually much greater than that of toluene. The attraction of ethanol to polypropylene ($\theta = 47°$) is much less than that of water to nylon ($\theta = 71°$). Only in comparisons involving a single liquid is an increase in attraction accompanied by a systematic decrease in θ.

TABLE 2

Water Wetting of Single Fibers and Filaments

Material	Work of adhesion (erg/cm^2)		Hysteresis Ratio	Fiber diam X 10^2 (cm)
	Advancing	Receding		
Cotton (scoured)	148.3	161.2	1.09	0.1
Glass	138.5	138.5	1.00	0.2
Carbon	128.0	148.2	1.16	0.5
Nylon	106.3	131.4	1.24	3.8
Polycarbonate	94.1	165.0	1.75	1.3
Polyester	84.9	108.6	1.28	2.5
Polystyrene	81.6	106.6	1.31	4.8
Polypropylene	77.4	115.4	1.49	2.5
Teflon	56.1	128.5	2.29	2.9
Wool	45.6	111.3	2.44	0.2

Investigators reporting and commenting on the previously mentioned hysteresis effect have considered it most likely the result of transient surface contamination or surface roughness. Quantifying degree of wetting in terms of θ or cos θ makes it awkward if not impossible to measure the magnitude of this hysteresis. However, work of adhesion values allow the calculation of a ratio $(W_{SL}^{\;r}/W_{SL}^{\;a})$, which is a clear indication of the extent to which a liquid is more attracted to a pre-wet surface than to a "dry" one. Table 2 lists hysteresis data for water with a variety of materials, showing that there is a general, inverse correlation between the advancing work of adhesion and the magnitude of hysteresis. In other words, if the original surface is not very attractive to a liquid, forced wetting (that is, imposed immersion) will leave it in a state that is much more attractive to the adjacent liquid. On the other hand, prewetting a glass fiber does not change its attraction to water, which is considerable to begin with. While the complete explanation of wetting hysteresis may pose quite a complex problem, anyone studying it would be well advised to use work of adhesion data rather than contact angles. Table 2 also includes the diameters of the fibers used, indicating the wide range that can be studied by the wetting-force technique.

The question of whether or not the highly curved surface of a fiber can influence its inherent polymer-surface wettability is one that has

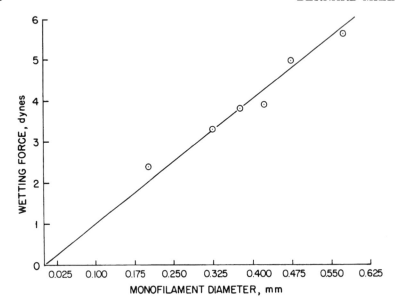

FIG. 10. Wetting forces of a set of nylon filaments with different
diameters. (Reprinted from Ref. 20, p. 362, by courtesy of Textile
Research Institute.)

been only sporadically discussed, most likely because reliable contact-
angle measurements were hard to come by. Wetting-force data on filaments of any diameter and shape can be easily obtained, and all the evidence available at this time indicates that such curvature has no effect
on wettability. For example, Figure 10 shows how wetting force increases linearly with diameter for a series of nylon filaments made
from the same polymer stock, or, in other words, that all these filaments had the same specific wettability.

IV. DYNAMIC WETTING

The static arrangement of a liquid in equilibrium contact with a
solid surface can be altered if either the liquid or solid are caused to
move by some external force. Most studies of this effect have been
performed by forcing liquids through capillary tubes [34]; the shape
of the advancing liquid-vapor interface was observed to be noticeably
sensitive to its velocity. More pertinent to filament wetting is the situation where a moving solid penetrates the surface of a liquid, passing
either from the dry side into the liquid or vice versa.

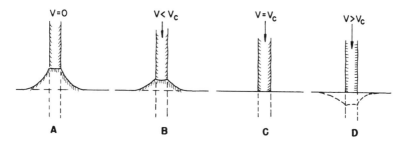

FIG. 11. Changes in the liquid meniscus around a vertical filament with downward motion of the filament. (Reprinted from Ref. 41,
p. 884, by courtesy of Textile Research Institute.)

When dry filament moves into a liquid the general effect on the
three-phase interface is shown in Figure 11. If there is positive wetting when the filament is stationary (A), there is a raised meniscus
and a contact angle less than 90°. As downward motion is imparted
to the fiber, the meniscus recedes and the contact angle becomes
larger (B). In general, it is possible to impose a velocity of the filament sufficient to eliminate the meniscus rise, or, in other words, to
bring the contact angle to 90° (C). A further increase in velocity may
produce an actual depression of the interface with a contact angle ex-
ceeding 90° (D). Even if the initial static condition does not show positive wetting, the same general trend of increasing contact angle with
increasing velocity can occur.

Only a few investigators have published data on this phenomenon.
Ellison and Tejada [35] used direct photographic observation of the
interface when a continuous length of nylon filament was drawn into a
liquid and reported cases where the increase in contact angle appeared
to reach a limit at less than 90° and did not follow the overall trend described in Figure 11. Inverarity carried out a more extensive study in
a similar manner with a variety of filaments [36]. His results revealed a general trend wherein the contact angle, while practically
constant at low velocities, increased systematically at higher velocities and in many cases nearly reached 180° (Fig. 12). Inverarity's
data showed that contact angle was dependent on the logarithm of the
product of liquid viscosity and filament velocity. These results were
interpreted to mean that viscous drag could be a dominant factor in the
formation of a dynamic contact angle. Schwartz and Tejada [37] measured dynamic contact angles, up to 90°, for a series of organic liquids
on moving filaments of stainless steel and polymers, 500 µm in diameter. They interpreted their results in terms of three different mechanisms governing the dependence of contact angle on filament velocity:

1. Initially the dynamic contact angle stays equal to the equilibrium
value as the velocity is increased (the Elliott-Riddiford [34] or Hansen-
Miotto [38] mode).

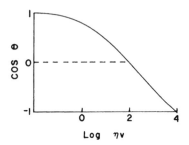

FIG. 12. General nature of the dependence of the dynamic contact angle on filament velocity and liquid viscosity.

2. In the next velocity range, the relation between contact angle and velocity is controlled by molecular motion along the filament surface; the retarding force is dependent on an interfacial viscosity (the Blake-Haynes [39] mode).

3. Finally, a surging mode of flow predominates, with the retarding force dependent primarily on the bulk viscosity (the Friz [40] mode when $\theta < 90°$).

A simpler experimental approach was used at Textile Research Institute by Dyba and Miller [41] to obtain a single-valued property for a given liquid-filament combination—the velocity required to produce a 90° contact angle (designated the rise-cancelling velocity V_c). The apparatus used for determining this velocity consisted of two major components: a mechanical drive system for advancing dry filament at a variable speed into a liquid held at a fixed temperature, and an optical system for observing the image of the intersection of the filament with the surface of the liquid. These are shown schematically in Figure 13. The filament drive included a trailing weight for maintaining tautness and a preheater to ensure that the filament was dry before contact. The optical system allowed light from a six-volt lamp to be reflected into the liquid bath by the first mirror at an angle sufficient to cause it to be totally reflected by the underside of the liquid surface back through the bath and out to the second mirror. The latter was placed so that it would reflect the beam through a projection lens which produced an enlarged image on a screen about 2 meters away. This arrangement makes use of the fact that undispersed reflection from the liquid surface will occur only if the surface is flat; any curvature will cause a portion of the light to be removed from the final image on the screen. Thus the curved rise of liquid about a filament appears as a dark ring about the central image of the filament against the bright image from the level liquid farther out. To perform the experiment, the drive system is used to gradually increase the velocity of the entering filament until the dark ring just disappears, indicating that the liquid surface is completely flat (a 90° contact angle). The velocity that is just sufficient to cause this is the rise-cancelling velocity V_c.

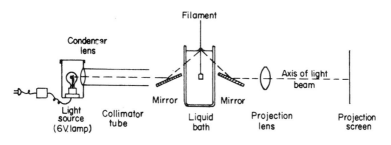

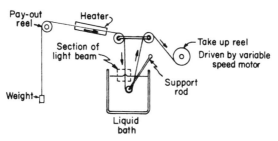

FIG. 13.   Apparatus for determining the rise-cancelling velocity. (Reprinted from Ref. 41, p. 885, by courtesy of Textile Research Institute.)

The above procedure is best suited for studying the effect of wetting-liquid properties, i. e. , viscosity or the presence of dissolved solutes or surfactants [42]. V_c does not seem to be very sensitive to the chemical nature of the filament; even such extremely different materials as nylon 66 and Teflon show just about the same response in a given liquid (see Fig. 14).   On the other hand, this property is significantly affected by both physical and chemical aspects of the wetting liquid.

The technique for obtaining V_c was taken up by Kimmel and Steiger to study what they called the wetting rate of cotton and rayon yarns as a function of water temperature [43].   They found evidence of an abrupt change in wetting behavior at certain temperatures which they attributed to the presence of surface waxes or finishes.

Rise-cancelling velocity measurements were used by Schick [44] as part of a study of the factors that can influence the friction induced by moving synthetic fibers.   Friction was observed to increase abruptly at a certain critical velocity and in nearly all cases this corresponded to the rise-cancelling velocity.   This observed relationship between dynamic

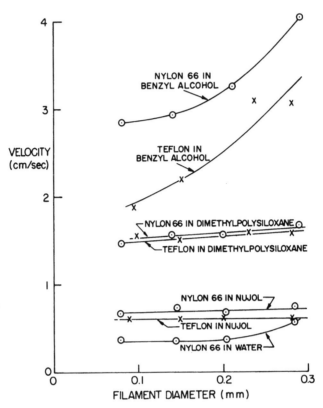

FIG. 14.   Rise-cancelling velocities of nylon 66 and Teflon filaments in various liquids. (Reprinted from Ref. 41, p. 887, by courtesy of Textile Research Institute.)

wetting behavior and friction is extremely interesting and raises the possibility that one of these two properties may be chiefly responsible for the other.

The fact that the dynamic contact angle does not change when the filament is advancing at a slow rate leads to a useful application using Wilhelmy wetting-force measurements on a moving filament. By arranging to record the wetting force continuously as the liquid-air interface moves up or down the filament, one can monitor random or systematic surface irregularities with considerable precision and sensitivity. Bendure [33] was the first to demonstrate this technique, using a polyester filament that most likely had titanium dioxide particles embedded in its surface. An example of the reproducibility of such a scan is shown in Figure 15.

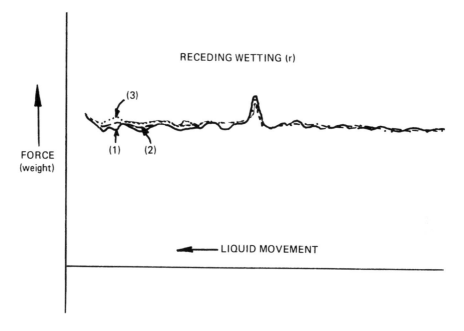

FIG. 15. Results of three continuous wetting-force scans over the same length of fiber surface.

Investigations of the reverse dynamic wetting situation, where wetted filament is drawn out of a liquid, have been mainly concerned with the equilibrium thickness of the film adhering to the emerging filament surface. Two theoretical approaches, one by Deryagin [45] (subsequently modified by White and Tallmadge [46]) and another by Carrol and Lucassen [47], differ in their conclusions. Deryagin predicts that film thickness will depend on the withdrawal speed raised to the two-thirds power, while the latter authors conclude that the exponent will be one-half. Nicol and Wilson [48] have carried out experiments using a variety of liquids in order to test these alternatives. They found clear-cut evidence supporting the Deryagin theory when pure liquids were used, but when surface-active agents were present the results indicated that even slight contamination could affect film thickness to a considerable degree.

In contrast, Roe [49] states that many liquids will not coat an emerging filament uniformly, even if they can spread spontaneously on the surface (that is, have an equilibrium contact angle of 0°). Instead, the liquid will form itself into discrete beads covering no more than three-fourths of the available surface. Roe has analyzed this effect as a special case of the general phenomenon of the instability of moving liquid columns.

ACKNOWLEDGMENT

This writer's knowledge of the subject of fiber wetting has been considerably enhanced as a result of his association with the following people: Dr. Raymond V. Dyba, Dr. Ian A. Brown, and Dr. Raymond A. Young. Preparation of this manuscript was greatly assisted by the advice and help of Dr. Harriet Heilweil.

REFERENCES

1. D. M. Brewis, Polym. Eng. Sci., 7, 17 (1967).
2. R. E. Johnson, Jr. and R. H. Dettre, J. Colloid Sci., 20, 173 (1965).
3. T. Young, Proc. Roy. Soc., December 1804, and Phil. Trans., 1805, p. 65; A. Dupré, "Théorie Mécanique de la Chaleur," Paris, 1869, p. 2883.
4. N. K. Adam, "The Physics and Chemistry of Surfaces," Third Edition, Dover, New York, 1968, p. 179.
5. A. W. Adamson, "Physical Chemistry of Surfaces," Interscience, New York, 1960, p. 355.
6. J. R. Huntsberger, Chem. Eng. News, 42 (44), 82 (1964).
7. D. H. Kaelble, J. Adhesion, 2, 66 (1969).
8. D. H. Kaelble, P. J. Dynes, and E. H. Cirlin, J. Adhesion, 6, 23 (1974).
9. D. H. Kaelble, P. J. Dynes, and L. Maus, J. Adhesion, 6, 239 (1974).
10. H. W. Fox and W. A. Zisman, J. Colloid Sci., 5, 514 (1950).
11. A. H. Ellison, H. W. Fox, and W. A. Zisman, J. Phys. Chem., 58, 503 (1954).
12. W. A. Zisman, Relationship of the Equilibrium Contact Angle to Liquid and Solid Construction, in "Contact Angle, Wettability and Adhesion," Advances in Chemistry Series, No. 43, American Chemical Society, Washington, D.C., 1964.
13. J. R. Dann, J. Colloid Interfac. Sci., 32, 302 (1970).
14. A. W. Neumann, Adv. Colloid Interfac. Sci., 4, 105 (1974).
15. F. M. Fowkes, J. Phys. Chem., 66, 382 (1962).
16. L. A. Girifalco and R. J. Good, J. Phys. Chem., 61, 904 (1957).
17. A. El-Shimi and E. D. Goddard, J. Colloid Interfac. Sci., 48, 242 (1974).
18. S. Wu, J. Polym. Sci., Part C, 34, 19 (1971).
19. F. E. Bartell and L. S. Bartell, J. Amer. Chem. Soc., 56, 2205 (1934).
20. B. Miller and R. A. Young, Text. Res. J., 45, 359 (1975).
21. S. H. Mason and C. F. Prutton, "Principles of Physical Chemistry," 3rd ed., MacMillan, New York, 1958, p. 226.
22. J. W. Bluhm, American Laboratory, November 1974, p. 31.

23. F. W. Minor, A. M. Schwartz, E. A. Wulkow, and L. C. Buckles, Text. Res. J., 29, 940 (1959).
24. A. M. Schwartz and C. A. Rader, Physics and Application of Surface Active Substance, Proc. 4th Intern. Congr. Chemistry, Vol. 2, 1964, p. 383.
25. J. -I. Yamaki and Y. Katayama, J. Appl. Polym. Sci., 19, 2897 (1975).
26. W. C. Jones and M. C. Porter, J. Colloid Interfac. Sci., 24, 1 (1967).
27. T. H. Grindstaff, Text. Res. J., 39, 958 (1969).
28. T. H. Grindstaff and H. T. Patterson, Text. Res. J., 45, 760 (1975).
29. J. Wilhelmy, Ann. Physik, 119, 177 (1863).
30. G. E. Collins, J. Text. Inst., 38, T73 (1947).
31. A. J. G. Allan, J. Colloid Sci., 13, 273 (1958).
32. A. M. Schwartz and F. W. Minor, J. Colloid Sci., 14, 572 (1959).
33. R. L. Bendure, J. Colloid Interfac. Sci., 42, 137 (1973).
34. G. E. P. Elliott and A. C. Riddiford, J. Colloid Interfac. Sci., 23, 389 (1967).
35. A. H. Ellison and J. B. Tejada, NASA Report #CR72441, 1968.
36. G. Inverarity, Brit. Polym. J., 1, 245 (1969).
37. A. M. Schwartz and S. B. Tejada, J. Colloid Interfac. Sci., 38, 359 (1972).
38. R. S. Hansen and M. Miotto, J. Amer. Chem. Soc., 79, 1765 (1957).
39. T. D. Blake and J. M. Haynes, J. Colloid Interfac. Sci., 30, 421 (1969).
40. G. Friz, Z. Angew. Phys., 19, 374 (1965).
41. R. V. Dyba and B. Miller, Text. Res. J., 40, 884 (1970).
42. R. V. Dyba and B. Miller, Text. Res. J., 41, 978 (1971).
43. J. M. Kimmel and F. H. Steiger, Ind. Eng. Chem., Prod. Res. Dev., 9, 259 (1970).
44. M. J. Schick, Text. Res. J., 44, 758 (1974).
45. B. M. Deryagin and S. M. Levi, "Film Coating Theory," Focal Press, London, 1964, pp. 46-48.
46. D. A. White and J. A. Tallmadge, Amer. Inst. Chem. Eng. J., 13, 745 (1967).
47. B. J. Carrol and J. Lucassen, Chem. Eng. Sci., 28, 23 (1973).
48. S. K. Nicol and G. R. Wilson, Aust. J. Chem., 28, 461 (1975).
49. Ryong-Joon Roe, J. Colloid Interfac. Sci., 50, 70 (1975).

Chapter 12

SOIL RELEASE BY TEXTILE SURFACES

H. T. Patterson and T. H. Grindstaff

Fiber Surface Research Section
Textile Fibers Department
E. I. du Pont de Nemours & Co.
Kinston, North Carolina

I. INTRODUCTION

Scientific investigation of soil release from textile structures has greatly proliferated in recent years, mainly because of the continually increasing use of man-made fibers, the introduction of ease-of-care and soil-release fabric coatings and the availability of a broad spectrum of detergent formulations. A comprehensive review of all the research in this field is beyond the scope of this short chapter. This survey will serve only as an introduction to most of the factors reported to influence soil release and will include some speculation on their interrelationships as well as references for further study. Techniques for quantifying soil pick-up and release will also be discussed. Although principle emphasis will be placed on soil removal, pick-up of soils by textiles will also be considered where the mechanisms of pick-up and release interact. Contributions of fiber and fabric characteristics to soil removal will be considered more extensively than those of fabric coatings, which are the subject of another chapter in this work.

II. PROBLEM ANALYSIS

The word "soiling" in connection with textile surfaces most frequently denotes the unwanted accumulation of oily and/or particulate materials on the surfaces or interior of fibrous structures. At times soils may diffuse into the fibers, but most often they adhere to fiber surfaces, become lodged in surface flaws (cracks or pits) or are trapped between fibers. The presence of foreign matter is considered undesirable when it is visually apparent because of changes produced in fabric color or luster. Other unpleasant changes may include odor production or alterations in tactile characteristics.

The extent of soil removal during textile cleaning is affected by the substrate, the soil, the cleaning method, and interactions among all three. Each of these areas has been extensively investigated by many workers. Broad studies have been made of the total problem, but most frequently attention has been focused on limited portions of the overall picture. Naturally, conditions which may optimize one aspect of problem solution can conflict with requirements in another area. The outline below summarizes the major variables affecting textile soil removal.

TEXTILE SUBSTRATE CHARACTERISTICS:

Chemical nature, roughness and hardness of fiber (or coating) surfaces
Shape of fiber cross section

Yarn structure and twist
Fabric construction
Swelling during cleaning

SOIL PROPERTIES:

Affinity for substrate
Solubility and/or dispersibility in cleaning medium
Particle size and geometry (solid soils)
Interaction between oily and particulate elements
Amount present

CLEANING METHOD:

Liquid medium (water or organic solvent)
Surfactant type and concentration in detergent
Type and concentration of detergent additives (suspending agents
    or "builders", enzymes and "brightening" agents)
Temperature
Cleaning time
Agitation energy input

To apply the scientific method to the understanding of soil release from textile surfaces, each of the above variables should be altered separately and soil removal analytically determined. If the principle objective of a study is to evaluate cleaning formulations, washing machine performance, or soil-release-agent efficiency, it may become desirable to employ use-soiled fabrics and visual evaluation by a panel of observers. This approach can reveal advantages or disadvantages important to the commercialization of new products more readily than a study involving better-controlled, quantitative determination of soil amount. However, basic understanding of soil removal developed by quantitative studies is needed for rapid and efficient design of new products and procedures to solve practical problems.

III. SOILED TEXTILE MODEL

The fundamental problem of soil removal from textiles most frequently involves separating oily and particulate matter from a complex fibrous structure and carrying it away in a fluid medium with minimum re-attachment to the fiber surfaces. Nine of the possible interfaces and junctions which may be involved are illustrated by sketches in Figure 1. An oil droplet which does not wet the fibers well (A) or a solid particle which touches lightly at only one point (B) can probably

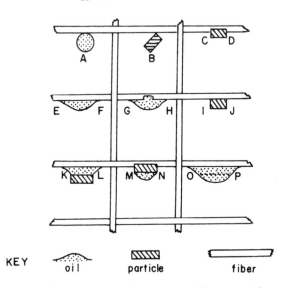

FIG. 1.   Fibrous structure soiling model.

be readily flushed or shaken from a fabric.   In contrast, a solid particle imbedded in the fiber surface (C-D), oil droplets wetting the fiber well (E-F), and oil droplets imbedded in fiber fissures (G-H) probably need vigorous mechanical and/or chemical cleaning action.   Soils may adhere to the fiber (E-F or I-J) or to other soiling materials (K-L or M-N).   In addition to the three different types of interfaces involving adhesion between different materials (E-F, I-J and K-L or M-N) the possibility for cohesional separation exists in which part of an oil droplet is removed by cleaning (O-P).   In extreme cases there may be rupture of solid particles or abrasion of the surfaces of the fibers which could lead to soil removal, but these events are considered unlikely or undesirable in general cleaning practices.

Complete soil removal is seldom if ever achieved in most commercial and home cleaning operations [1].   As little as 0.1 to 0.2 weight percent of an oily soil would be sufficient to provide a thin layer, possibly a monolayer, covering most of the fiber surfaces, particularly on the relatively smooth man-made fibers.   This suggests that cohesional separation (O-P) of oil deposits represents the more probable type of oil separation and that rupture at the oil-fiber interface is not required to provide relatively "clean" textiles.   It also implies that the importance of variables affecting the percentage of soil removal is modified by the amount of soil present.   Removal of 95% of a large amount of oily soil may be considered highly effective but still leave a coating on nearly all the fiber surfaces.   At low soil levels fiber-soil separation must be highly effective to produce analytically high efficiency ratings.

IV. SOIL VARIATIONS

A. Composition of Natural Soils

Any foreign material which adversely affects the color, odor, or tactile aesthetics of textile materials has been called a "soil" [2]. The composition of soils naturally varies with the uses to which textiles are put (clothing, drapes, carpets) and the environmental conditions during their use (industrial smog vs. clean rural air; dry, hot, dusty areas vs. damp, cool, muddy areas). Analyses by many workers [3, 4, 5] have shown that clothing in contact with human skin picks up measurable amounts of greasy human sebum as a soil. Brown [3] found that combined carbon tetrachloride extracts of shirts, socks, pillowcases and towels contained about 31% free fatty acids, 29% triglycerides, 15% fatty alcohols and cholestrol, 21% hydrocarbons, and 3.3% short-chain fats and oils. Although analyses vary somewhat among different workers, Brown's results provide a reasonable general guide to the type and proportion of ingredients for use in simulating soil composed primarily of body fats and oils.

Bowers and Chantry [4] extracted an average of 1.2 g. of sebum from T-shirts worn for only one day. This amount represented about 1% based on garmet weight (ca. 100 g.) and indicated a need to develop sensitive analytical techniques for quantitative soil removal studies.

Powe [5] has pointed out that skin particles are a form of soil found on clothing areas where abrasion occurs between fabric and skin. These particles are groups of epithelial cells averaging $25\text{-}30\mu$ in diameter which, because of their flexibility, conform closely to fiber surfaces and are quite difficult to remove. Although this source of soiling material would reasonably be expected to influence fabric cleaning extensively, it is seldom, if ever, considered in studies involving synthetic soils.

Dust deposited from the air makes up a large part of the solid particulate soil found on outer garments, drapes, carpets and other textiles. It is reasonable to assume that this material is similar to the dust found in city streets. Analyses of street dirt from six large cities by Sanders and Lambert [6] showed a marked uniformity of inorganic constituents. Levels of ether-soluble organic material uncovered a variation of from 4.9% in Detroit to 12.8% in St. Louis.

Over 50% of the particles in the street soil analyzed by Sanders and Lambert was in the 0 to 4μ size range [6]. Jones [7] reported that particles below about 0.2μ are nearly impossible to remove from textiles by laundering, probably because of their high surface-to-volume ratio and the ease with which they can become lodged in small surface irregularities. Even particles as large as 5μ can cause problems. This dependence of removability on particle size demands that strict attention be given to this variable in the preparation of synthetic soils.

Although materials picked up from human skin and dust deposited from the air are major components in routine textile soiling, a cleaning formula, soil-release agent, or washing machine designed by laboratory studies to remove these materials alone could be an unexpected failure in commercial use. Some of the most difficult soils to remove are stains, which are usually concentrated in localized areas on textiles and vary widely in composition depending on their source. These can include such diverse materials as cooking grease, used motor oils, blood, grass stains, ink, paint, and food residues. These soil types generally require special treatments for removal such as spotting with organic solvents, sorption by finely divided solids, or exposure to high concentrations of liquid detergents. More drastic actions include oxidation to more readily removed polar compounds by bleaching agents and decomposition to lower molecular weight materials by enzyme action.

B. Synthetic Soils

A review of the literature in a recent publication by Byrne [8] indicated that many believe artificial soils do not simulate natural soils. Among the potential problems presented is that the artificial soils are frequently applied all at once at too high levels, whereas realistic soiling takes place by repeated application at low levels. Despite these possibly valid objections, much research on removing soil from textiles is carried out by laboratory application of soil mixtures selected to represent the major constituents of natural soils. Lack of precise control over the level and type of compounds adsorbed and the long time frequently required for accumulation of analyzable levels are obvious drawbacks to employing use-soiling techniques. The technique to be used to establish the degree of soiling before and after cleaning often determines the technique selected. A procedure based on light reflectance does not require the control over soil chemistry required by techniques based on amounts of soil present. The highly sensitive radio tracer technique is often used to follow tagged synthetic soil ingredients.

1. Ingredient Selection

The obvious complexity of diverse mixtures of natural materials makes their precise duplication in controlled studies essentially impossible. Efforts to establish the effectiveness of various synthetic soils fill the scientific literature with reports, each covering a different, complex mixture of natural products and pure chemicals. Most of these formulations include one or more of the following:

PARTICULATE SOLIDS:

Clays and silicates
Carbon black and graphite
Metal oxides
Natural dust (vacuum cleaner dirt)

OILS AND FATTY SOLIDS:

Aliphatic acids (saturated and unsaturated)
Alcohols (aliphatic and cholesterol)
Hydrocarbons (pure and partly oxidized - used motor oil)
Triglycerides (saturated and unsaturated)
Natural products (olive oil, tallow)

PROTEINS AND CARBOHYDRATES:

Vegetable matter
Food products

MISCELLANEOUS:

Dyes (natural and synthetic)
Cosmetics

For generalized comparisons of soil removal, tests have been developed using standardized mixtures of soils such as that proposed by Florio and Mersereau [9]. In contrast, basic studies of soil removal mechanisms have usually been performed with chemically pure materials and mixtures. The majority of studies includes one or more particulate materials mixed with one or more liquid products.

Selection of the solid components is particularly difficult because wide varieties of particle sizes, shapes and surface conditions are available for each chemical type. Carbon black has been used widely as a synthetic soil because a small amount can produce extensive greying of textiles similar to that noted in natural soiling. However, in a recent study of soil redeposition in laundering [10] not one of nine carbon samples produced fabric greying which corresponded to coloration produced by repeated laundering. Suitable clay samples are also difficult to obtain. Representative oily soil compositions are more easily made up from pure chemical ingredients. It is important to adhere to the proportions of polar and non-polar and low- and high-molecular-weight components of natural oily soils because of the influence these variables can have on removal.

Fort, et al [11] pointed out the importance of oily soil constituents in causing solid soils to adhere to fiber surfaces. Precise duplication of natural soiling would thus require both solid and oily components in a synthetic mixture, since oils are found almost universally as surface contaminants. Kissa [12] recently concluded that a simple model soil (i. e. iron oxide) could be used instead of natural soiling to obtain relative soil resistance values. He found that amounts of solid synthetic and natural soils on various fabrics correlate well when oil or fat is absent.

An infrequently mentioned variable affecting performance of oil-solid soil mixtures is the extent of interaction between the liquid and particulate components. For example, an equal mixture of used motor oil and vacuum cleaner dirt (VCD) formed a plastic, semi-solid mass [13]. A similar mixture of oil and clay was a pourable fluid even though the average clay-particle size was one percent that of the VCD. A polyester fabric retained up to four times more oil-VCD mixture than oil-clay mixture after a laboratory laundering test. This greater retention of the oil-VCD is evident in the micrographs in Figure 2. (Note: marks on the fiber surfaces in the lower picture were produced by rubbing the applied soils to imbed them in the fabric.)

Martin and Fulton [14] suggest that synthetic soil mixtures should be sufficiently difficult to remove that some portion remains after cleaning. In this respect the synthetic soils would duplicate the most common experience in practical cleaning operations. These workers concluded that a synthetic soil is suitable if the rate equation for removal differs only in slope from that for natural soils. This is illustrated in Figure 3, which indicates that in a practical cleaning time, three different detergents removed all of a natural soil. Small errors in cleaning time would cause large errors in the amount removed, which would influence judgment of detergent effectiveness. Removal plots for the synthetic soils flattened out to provide a much less equivocal evaluation of the detergents. Although these relative results might be acceptable from a practical standpoint, other factors would need consideration to confirm soil removal mechanisms. Less than complete soil removal may represent removal only from capillary spaces between fibers in a fabric, but an oil monolayer may remain on all fiber surfaces. This last small fraction of soil is most difficult to detect, as is the uniformity of soil distribution in a laundered sample. Caution is needed in the application of surface-adsorption interpretations to soil removal theory.

The usefulness of synthetic oily soils in research studies has been augmented by radio-tagging of individual components of mixtures. In

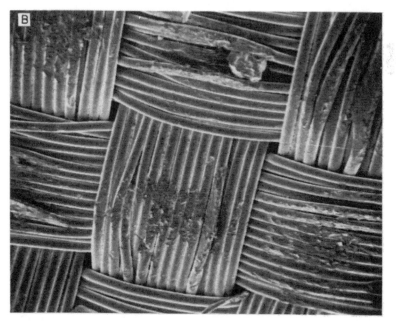

FIG. 2. Comparison of removal of oil-particulate soil mixtures.
(A) Oil-vacuum cleaner dirt. (B) Oil-clay.

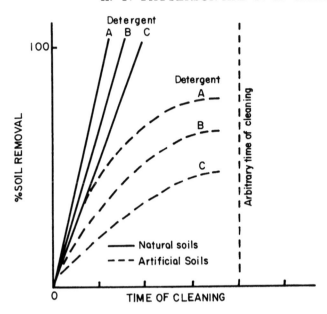

FIG. 3.   Effect of natural and artificial soils on evaluating detergency. Reprinted from Ref. 12, p. 22, by courtesy of John Wiley & Sons, Inc.

this way Fort, et al [15] were able to establish the order of increase in desorption rates of four oily soil components as: hydrocarbon < triglyceride < fatty alcohol < fatty acid.   The percentage of oily soil removed in a given time in this work was apparently independent of the amount of soil initially present.   They also showed that the rate of deterging of one constituent in a mixture was increased by the presence of more readily removed materials and decreased by compounds more difficult to remove.   Carbon-14-labelled amorphous carbon was used in detergency studies by Grindstaff, et al [16] to show that hydrophilic carbon was more readily removed from polymer surfaces than hydrophobic carbon.   Mixing carbon with a synthetic sebum mixture had little or no effect on its removability with several fabric-detergent combinations.

2. Soil Application

A key factor influencing the removal characteristics of synthetic soil mixtures is the procedure used for their application.   Some of the variables involved include:

METHOD

Contact with 100% soil
Application of soil solutions or dispersions

Tumbling with particulate soils
Transfer from soiled samples

LEVEL

Single treatment
Multiple treatment

AGING

ABRASION

Two potential problems with application of 100% materials are the
difficulty of obtaining uniform application over a large area and the
generally unavoidable high level of loading. In studies involving removal of stains the synthetic conditions employed essentially duplicate
textile-use situations [17]. In another form of 100% soil application,
test pieces have been hung in suspensions of air-borne dust using a
laboratory apparatus developed for this purpose by Rees [18]. Un-
diluted oils have been applied to fabrics by Kissa [19], who sucked
them into fabric samples in a filtration process and pressed out excess liquid.

The most commonly used soiling procedure is a single application of dilute solutions or suspensions of the soils in volatile fluids.
Some exceptions taken to the validity of this procedure include the
possible swelling of fibers by organic solvents. In the case of cotton
it has been claimed [20] that the internal structure of the fiber is
opened by solvents used in applying oily soils. This could permit
penetration and entrapment of the soils and hinder their subsequent
removal. Investigation of this possibility showed [21] that removal
of radio-tagged triolein from cotton fabric was the same when similar levels were applied from a solution in carbon tetrachloride or by
migration of the 100% liquid from other soiled fabrics used to sand-
wich the test piece.

Tumbling dry particulate soils in a drum lined with test materials is frequently used in tests of carpet materials [22]. Heavy balls
or metal jacks with balls on the prongs (tetrapod walkers) are added
to simulate natural grinding of soils into the fabrics by walking. In
another form of tumbling application, felt cubes saturated with soils
are used to accomplish uniform soil transfer to test fabrics [23].
Kissa [24] developed a procedure for applying dry soils with the Acceleroter abrasion tester and showd its correlation with soiling values
obtained by other dynamic methods. The abrasive lining normally used
in the Acceleroter was removed and measured amounts of soil were
added to the chamber before rotating fabric samples for controlled
times at varying velocities.

Transfer of soils from soiled to clean fabrics can be carried out by pressing the clean and soiled samples together [21] or by cleaning them together as in the case of redeposition studies [25]. A special case of transfer involves "printing" of soils onto test fabrics [26].

One procedure which eliminates the complications of yarn and fabric structures is to soil finely chopped (~1 mm. long) fibers with soil solutions or dispersions [27]. Another simplifying procedure is to spread soils on film samples corresponding to the fiber types being studied [15]. Both these techniques are more suitable for research studies than for duplication of natural soiling conditions.

Abrasion of samples during soil applications is frequently carried out to match conditions of natural soil accumulation more closely. This occurs in tumbling type applications, and in some cases test pieces may actually be rubbed to grind in the soil [28]. Fiber characteristics and fabric construction are important substrate variables affecting this operation. Surface damage can increase the available fiber area for sorption and provide crevices for mechanical entrapment. The heat generated may promote soil diffusion into filaments, and deeper soil deposition in the test piece interior can also be attained. From a practical standpoint abrasion of applied soil is undoubtedly important, but the multiplicity of only partially controlled conditions makes separation of variables in a fundamental study quite difficult.

C. Soil Detection

The procedures for establishing soil accumulation and removal may be divided into the major categories of reflection and quantitative analysis. Most studies related to practical demonstrations of improved soil removal rely on reflectance procedures, since these duplicate most closely the visual appraisal of the success of the improvements.

If it is assumed that light reflection decreases linearly with increased amounts of soil present from the level with an unsoiled textile sample to the level after soiling, and that removal increases reflectance linearly, a relationship for % soil removal can be written [29]

$$D = \frac{100 \, (R_{wf} - R_{sf})}{(R_{of} - R_{sf})}$$

where

D = % soil removed

R_{wf} = reflection after cleaning

R_{of} = initial reflection

R_{sf} = soiled reflection.

Since reflectance can be assumed linearly porportional to soil level only over narrow soiling ranges, this relationship becomes more incorrect as the changes in reflectance increase. Bacon and Smith [30] applied one form of the Kubelka-Munk relationship [31] to improve the situation. Reflectivity results may be used in a form to help correct for scattering of light by the fabric and soil. Thus, instead of R values, the form $\left[\frac{(1-R)^2}{2R}\right]$ is used to derive the equation:

$$D = 100 \left[\frac{\frac{(1-R_{wf})^2}{2R_{wf}} - \frac{(1-R_{sf})^2}{2R_{sf}}}{\frac{(1-R_{of})^2}{2R_{of}} - \frac{(1-R_{sf})^2}{2R_{sf}}}\right]$$

Variables such as textile and soil colors [32], the wave length of incident light [33], the presence of fluorescent fabric brighteners [34], and textile construction must still be considered.

Quantitative soil analyses have been accomplished by chemical [35], gravimetric [19], and radiotracer [15, 36] techniques. Special procedures such as X-ray fluorescence [37], neutron activation [38], and differential calorimetry [39] have also been tried or suggested. Each technique is subject to certain drawbacks such as use of non-representative synthetic soils, high soil levels needed for detection, difficulties of introducing tagged elements, and many more. Appropriate attention to the many conflicting variables important to soiling can make any of them useful for increasing knowledge of soil removal mechanisms. Kissa [35] demonstrated a correlation between reflectance soiling values calculated by the Kubelka-Munk equation and those determined chemically using iron oxide as a soil.

Frequent use has been made of tracer methods in which a material with distinctive optical or other readily detectable characteristic is added to a soil mixture to detect changes in the total amount of material present [40]. The obvious difficulties arising from possible differences in absorbance and removal characteristics between the tracer and the bulk of the soil must be eliminated to make this procedure successful.

Cramer [41] has published an extensive review of evaluation techniques.

V. TEXTILE SURFACES

A. Chemical Composition

Current textile soiling studies must consider not only the chemical constitution of the fibers in a sample, but also the composition of specialty coatings which may be present. With the advent of permanent press, soil-release finishes, flame retardants, permanent antistats, and many other fabric treatments, it is necessary to know the complete history of a fabric sample in order to establish details of its surface composition.

As indicated in Table 1, surfaces of both fibers and their coatings represent a wide variety of organic polymers which include almost any stable organic grouping. A convenient approach to classification of these materials in a manner related to their interaction with soils is on the basis of their surface energy. The ease with which soils are accumulated by textile surfaces and the resistance offered to soil removal depend to a large extent on the energy released by formation of interfaces in Figure 1 such as E-F and I-J (between soils and fiber surface) and M-N (between liquid and solid soils). Polymers important to textiles are classified as having low surface energy (ranging from about 10 to 50 ergs/cm^2 as indicated in Table 1). This places them in a class distinctly different from inorganic surfaces which may range from 290 ergs/cm^2 for glass [44] to 9820 ergs/cm^2 for the 100 face of diamond [45]. However, the relatively narrow 40-erg spread for textile polymers is sufficient to produce major variations in soiling behavior and is logically controlled by the chemical constitution of the surfaces involved.

Recent research on the surface energy of polymers has made extensive use of measurements of contact angles of liquids with varying surface tensions and chemical constitution [42]. One result of these studies is the development of the concept that the total energy of a polymer surface may be roughly divided into polar and nonpolar effects depending on the molecular interactions involved. Nonpolar interactions [46] have been mainly identified with Van der Waals or dispersion forces which are sufficient to form strong adhesive bonds even between hydrocarbon chains.

The polar portion of the surface energy of polymers indicated in Table 1 has been associated mainly with hydrogen-bond formation between the interacting surfaces [47]. The presence of ether, amide, hydroxyl, carboxyl, sulfonate and other polar groups in the surfaces of textiles can thus increase their ability to bind soil molecules. The many hydroxyl and ether groups in the surface of a cellulosic fiber contribute to its having a higher surface energy than a polyethylene fiber and may thus promote the greater accumulation of surface soil. The hydrogen bonds involved, however, are readily disrupted by water and/or surfactants. Morover, water may approach the fiber-soil interface from inside a cellulosic fiber because of its high moisture regain. Soil adhesion may thus be reduced even if water cannot readily permeate the soil layer. Since nonpolar as well as polar bonds are involved in soil adhesion and since other factors such as fabric geometry and soil suspension are involved, water alone cannot completely clean even soiled cellulosic fibers.

A study which clearly demonstrated the progressive change of wetting of polymer surfaces by systematically changing chemical constitution was carried out by Fort [48]. Surface wettability by water and thus the hydrogen bonding capability of a series of polyamides improved as the number of amide sites in the surface was increased by shortening the hydrocarbon groups linking them together. Two curves relating wettability improvement to decreased spacing between amide sites could be drawn (Fig. 4) because of the different molecular packing with odd and even numbers of carbon atoms between the polar sites. This observation illustrated the great sensitivity of wettability techniques to even small adjustments in surface composition.

A similar study with linear aliphatic polyester surfaces [49] showed improvement in wetting with increased concentration of ester sites (Fig. 4). Wettability for a given degree of modification in this series was poorer than with the nylons because of the relatively poorer wettability of the ester sites. Aliphatic polyesters are not used for fiber production, probably because of their low melting points. Water wettability of the widely used fiber polymer, polyethylene terephthalate, is higher than that of an equivalent aliphatic polyester probably because of the aromatic rings, but is still less than that of an equivalent polyamide. All the aliphatic polyesters in this study [49] were prepared with even numbers of carbon atoms in the monomers and thus did not show the odd-even fluctuations noted by Fort.

In the removal of oily soil from fiber surfaces by laundering, the wetting behavior involves not only water-fiber-air interactions but also water-fiber-soil relationships. In a study by Stewart and Whewell [50], contact angles of mineral oil with fibers were measured under water. This provided data indicating (Fig. 5) that contact angles of oil with polymer surfaces under water increased with increased water wettability of the fibers, even with no surfactant. When a surfactant was added, oil contact angles under water were markedly increased for all but the most nonpolar fibers. With a contact angle of over 140°, it is reasonable to expect that oil droplets would be easily removed by mechanical action during laundering.

Berch et al [51] related the influence of chemical constitution of fiber surface coatings to soil removability by anionic surfactants. In their work the receding contact angle of mineral oil on coating surfaces under aqueous detergent solution progressively increased as silicone, fluorocarbon and acrylic materials were used. Soil removal results for cellulose fabrics coated with these polymers increased directly with the increase in contact angle of mineral oil under surfactant solution (Fig. 6).

This complete inversion of fiber wetting by oily materials in air and in water has led to some confusion in the development of fiber coatings which will both repel soil during use and release it readily during laundering. Reduced soil accumulation in air is achieved with

TABLE 1

Representative Formulas and Surface Properties of Some Polymeric
Materials and Coatings Used in Textiles[a,b]

BASE FIBER	SURFACE ENERGY (ERGS/CM^2)			WATER CONTACT ANGLE (°)	STRUCTURE
	TOTAL	NONPOLAR	POLAR		
CELLULOSE	-	-	-	40	
POLYESTER	39.5	36.6	2.9	75	
ACRYLIC	-	-	-	48	R = CN OR CN & Cl MIXTURES
NYLON	41.5	33.7	7.8	70	
POLYTHENES	32.4	31.3	1.1	92	R = H OR ALKYL GROUPS

WOOL $\left[-NH-CH-CO-NH-\underset{R_2}{CH}\ CO-NH-\underset{R_3}{CH}-CO-\right]_x$ R_1, R_2 & R_3 = DIFFERENT GROUPS	61	-	-	-
FIBER COATING				
ACRYLIC ACID COPOLYMERS $\left[(-CH_2-\underset{COOH}{\overset{R}{C}}-)_x(-CH-\underset{C=O}{\overset{R}{C}}-)_y\right]$ $R=H$ OR CH_3 $R'=ALKYL$ GROUPS	-	-	-	-
POLYETHYLENE OXIDE $R(CH_2CO)_nR'$ $R'-O$, R & R' = END GROUPS OR ANCHORING SITES	-	-	-	43
FLUOROPOLYMERS $\left[-CH_2\ \underset{O=C-OR}{\overset{CH_3}{C}}-\right]_x$ $R = C_7F_{15}$	-	0.5	10.0	10.5
POLYSILOXANES $R\left[-\underset{CH_3}{\overset{CH_3}{Si}}-O-\underset{CH_3}{\overset{CH_3}{Si}}-O-\right]_x R'$ R & R' = END GROUPS OR ANCHORING SITES	-	1.6	20.5	22.1

aSurface Energy Data - [42] [43] [47]
bContact Angle Data - [50]

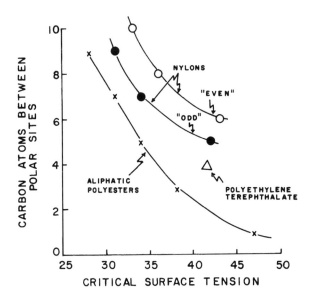

FIG. 4. Effect of polar site type and concentration on polymer critical surface tension for wetting. Data from [48] and [49].

polymer coatings having low surface energies such as silicones and fluorocarbons. These can repel aqueous-based soils and many oily materials [52, 53, 54, 55] but, when they do become soiled they are difficult to clean. Cleaning is a particular problem because many of the coatings are soft and permeable to the diffusion of stains into them during aging.

A workable compromise for these conflicting effects was developed by Sherman et al [56] by combining hydrophobic and hydrophilic polymer segments in a single durable fabric coating. Perfluoroaliphatic groups were used to repel stains in air, and polyethylene oxide segments provided soil-release hydrophilic sites in water. These workers explain the behavior of their hydrophobic-hydrophilic block copolymer finishes in terms of attainment of the lowest interfacial energy state for the environment in which they are placed. In air, the hydrophilic polyether groups are collapsed below the surface and a low-energy, soil-repelling perfluoroaliphatic surface is produced. In water, the polyether groups swell to provide a surface with low interfacial energy in the aqueous environment which facilitates soil removal.

Another fiber characteristic affected by chemical composition is the electrical charge on the surface which interacts with the

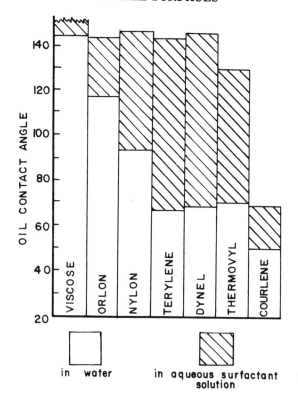

FIG. 5. Effect of surfactant on oil/fiber contact angle. Data from [50].

charges on soil particles [57]. Electrical forces can attract soil particles to produce unwanted soiling and hold the soil to make removal difficult. Surfaces of most fibers are negatively charged, but reports of their charge variations with chemical composition differ widely with different investigators [58, 59]. This may be explained by a considerable dependence of results on small differences in the environments used for the measurements. Data from Jacobasch [60] in Table 2 for distilled water indicate relatively lower potentials for fibers with extensive hydrogen-bonding capabilities such as cellulose, glass and nylon. Fluctuations in solution pH and varying concentrations of salts and surfactants modify charges on both fibers and soils.

The major dependence of soil removal on the zeta potential of fiber surfaces is in dispute; however, Jacobasch [60] reported an increase in difficulty of removal with increased fiber potential as shown

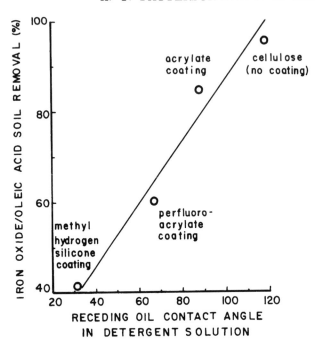

FIG. 6. Effect of oil wettability in detergent solution on soil removal from fabric coatings. Data from [51].

TABLE 2

Zeta-Potential of Textile Fibers[a]

Fiber	Potential (mV)[b]
Viscose rayon	22
Glass	28
6-nylon	33
Cellulose triacetate	37
Polyester	52
Polyvinyl chloride	55[c]
Polyacrylonitrite	60[c]

[a]From [60].
[b]Negative potential in distilled water.
[c]Approximate values because of poor reproducibility.

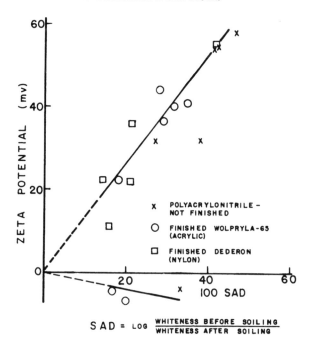

FIG. 7. Relationship between zeta potential and soiling. Reprinted from Ref. 60, p. 344, by courtesy of Faserforschung und Textil-
technik.

in Figure 7. To reach this conclusion he considered separately the effects of negative (most of the results) and positive potentials (the few
points below zero on the abscissa). This subject will be discussed further under detergents.

B. Fiber Physical Characteristics

1. Geometry

Extensive recent studies with scanning electron microscopy (see
also Chap. 7 of Part I) have established in fine detail the physical
characteristics of natural and man-made fibers. The scales on wool
fibers and the curled ribbon-like form of cotton fibers strongly suggest more extensive physical opportunities for both particulate and
oily soils to accumulate on them than on smooth, round fibers such
as some polyesters and nylons. In soil removal studies from flat
films [16], particulate carbon soil was found easier to wash from
cellulose than from nylon or polyester substrates (Table 3). In

TABLE 3

Effect of Polymer Substrates on Carbon Black Removal[a]

| | % Carbon black removed by washing | |
Substrate polymer	Film form[b]	Fabric form[c]
Cellulose	80	9
66-Nylon	35	79
Polyester	48	66

[a]From [16].
[b]0.01M sodium lauryl sulfate, 15 min, 60°C.
[c]0.33% built anionic detergent, 15 min, 60°C.

contrast, this particulate material was more difficult to wash from cotton fabric than from nylon or polyester fabrics. Fabric construction as well as fiber morphology probably affected these results and will be discussed below.

Compton and Hart [61, 62, 63], who deposited carbon black on cotton from water dispersions, concluded that geometric bonding was the principle mechanism affecting particulate soiling in the absence of oils and greases. Two kinds of geometric bonding were proposed: (1) micro-occlusion, in which fine particules are trapped in fiber surface crevices, and (2) macro-occlusion, in which soil particles are trapped in fabric interstices. These workers found that micro-occlusion was more important for rougher surfaced natural fibers than for smooth, round man-made fibers. They also observed the expected dependence of soiling on the relationship between the size of the particles and the size of the surface crevices into which they could fit.

Studies by Tsuzuki and Yabuuchi [64] indicated that fiber cross-sections affect the removal of oily as well as particulate soils. Removal of SAE-20 oil from fabrics of round, polyester filaments was less difficult than from similar fabrics with fibers having irregular cross-sections. Oil trapped in cleaned cotton fabrics was found in fiber irregularities by microscopic examination.

Roughness of polymer surfaces has long been recognized as an important variable affecting wetting by liquids. Differences between advancing and receding contact angles have been attributed to the retention of liquids in surface irregularities as contacting liquid is withdrawn. Variations in wettability as well as in soil entrapment produced by changes in fiber surface roughness must both be considered important to soil removal processes. Grindstaff [65] showed that the advancing contact angle between polyester fibers and thiodiglycol was decreased from $43 \pm 2.4°$ to $31 \pm 4.2°$ by a relative roughness increase of only 14%. The degree to which increased soil entrapment and modified wettability may compensate for each other in a roughened fiber is a matter for speculation.

2. Hardness

Hard soil particles imbedded in the surface of fibers (interface C-D in Fig. 1) are difficult to remove. The degree to which particles become imbedded under equivalent external forces is logically dependent on the hardness of the fibers. Trip and others [66, 67] examined the extent to which variations in the hardness of fiber coatings modified retention of particulate soils. Hard finishes coated most of the crevices in cotton fibers and reduced soil retention. Soft finishes also filled surface irregularities, but soiling was increased because the particles were imbedded in the coatings. Hard finishes must also be tough to resist cracking, since fissures in coatings can provide sites for soil entrapment similar to those in fiber surfaces [61, 62, 63].

3. Absorbtivity

In addition to the possibilities for geometric entrapment of soils in the surfaces of textile assemblies, some of the oily components can become molecularly entangled in the body of the filaments by diffusion. Variables affecting this process include the size and chemical nature of the soil molecules, the type of fiber and the time and temperature of interaction. Obviously the deeper the soils have diffused below the fiber surface, the more difficult it will be to remove them.

A useful procedure for making pertinent observations on this problem was demonstrated by Fort et al [15]. Carbon-14-labelled components similar to materials in human sebum were applied to one side of cellulose, nylon, or polyester films and the activity was measured on both sides. Since the films acted as beta-ray shields the count was naturally lower on the untreated side. When samples were heated and aged to promote diffusion, the count differential decreased as the tagged synthetic-soil components migrated into the films. In this way it was shown that octadecyl alcohol and octadecane diffused extensively into all three polymers at $80°C$, and that octadecane diffused more rapidly. Glyceryl tristearate showed little or no diffusion. Thus, diffusion rate varied inversely with increased polarity and size of the synthetic soil molecules. These studies also indicated that diffusion for octadecane decreased in the following order: cellulose > nylon > polyester.

The effect of diffusion on soil removal was demonstrated with polyester film and radio-tagged stearic acid. Removal of stearic acid by a nonionic detergent was essentially 100% after 15 minutes

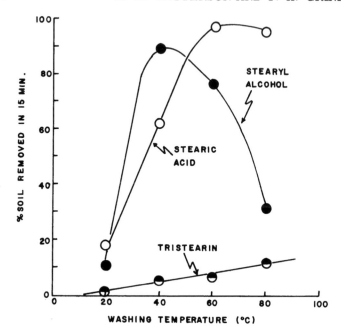

FIG. 8.   Effect of temperature on soil removal from polyester film
by 0.01M sodium lauryl sulfate.   Data from [15].

of washing; one minute of heating at 80°C before washing reduced removal to 70% and 10 minutes of heating reduced it to 30%.   This suggests that when soil removal is incomplete after laundering, the heat
employed in ironing can drive soils deeper into the fabric.   Elimination of the need for hot pressing in wash-wear fabrics can thus contribute to better cleanability.

Fort also found that increasing the temperature for removing oily
soil from polyester films increased the amount of soil removed up to
a maximum value and that higher temperatures could cause a reduction in removal.   As shown in Figure 8, the effect was most evident
with stearyl alcohol, less apparent with stearic acid, and not evident
with tristearin (except after long washing times).   A diffusion mechanism was confirmed with radio-tracer measurements and by bound-
ary-friction data.   The possibility for diffusion suggests that soiled
fabrics should be cleaned as soon as possible to reduce the opportunity for the "setting in" of soils.

Evidence for diffusion of oily materials into textile finishing coatings was obtained by Berch et al [51].   The contact angle of a paraffin oil on a polyacrylate film remained constant at 125° from 20 to 70°C,
but the contact angle for oleic acid decreased from about 100° to 50° in

the same temperature range. These authors concluded that the polar material had an increasing tendency to interact and dissolve in the film surface as the temperature increased.

Repeated soil-wash cycles with natural cellulosic fibers can result in accumulation of oily materials such as observed for cotton fibers by Oldenroth [1]. This result indicates the possibility of molecular diffusion as well as trapping of imbibed soils in the complex internal capillary structure of the fibers.

C. Yarn and Fabric Geometry

The structure of textile materials varies with the end uses to which they are put. Fibers may be loosely or tightly twisted into yarns and then converted into loose or tight fabrics depending on the warmth, cover, weight or tactile aesthetics desired [68, 69]. Variations in spacing between individual fibers and between fiber bundles control the extent to which internal surfaces contact soils or cleaning fluids. Tight twisting of yarn bundles or tight packing into fabrics can keep soils away from interior surfaces, but it can also prevent cleaning fluids from removing soils which do penetrate.

In a study of twist on soil pick-up by continuous filament acetate yarns Weatherburn and Baley [70] showed that soil retention increased with increased twist to a maximum level and then decreased with further twisting. By dissolving the fibers away from soiled yarns they showed that for up to twenty turns per inch particulate soils penetrated the interior of yarn bundles, but at sixty turns they were only on the surface.

A study of oily soils on polyester fabrics by Brown et al [71] showed that fabrics from staple fibers accumulated about five times more oily soil than those from continuous filament yarns under the same conditions. Moreover, removal of the oils was about twice as difficult from the staple fabrics. These differences correlated with the surface areas of the yarn bundles rather than with the areas of the individual filaments involved. Electron micrographs of cleaned staple fabrics showed that soil was trapped at crossover points of the filament pairs.

The same higher level of soil retention by fabrics from staple polyester yarn was demonstrated for particulate carbon soils by Grindstaff et al [16]. Roughly twice as much hydrophobic carbon was removed by laundering from a continuous filament fabric as from a staple material.

Bowers and Chantry [4] reported that less soil was retained after laundering by a low-twist, loosely woven cotton oxford cloth than by highly twisted, tightly packed broadcloth even though the latter was lighter in weight. Through statistical analyses they also established that fabric construction was only slightly less important than fiber type in controlling soil removal from cotton and polyester-cotton blend fabrics. Their application of a hydrophilic coating to polyester staple produced a 66% reduction in soil retention after washing when evaluations were carried out with loose staple, but this advantage was lost when the staple was converted into fabric. They speculated that reduced exposure of the hydrophilic surface in twisted yarn and greater soil pick-up by the higher-energy surface caused these results.

VI. CLEANING METHODS

A. Introduction

Of the three procedures widely used for removing soils from textile surfaces, laundering (aqueous), dry cleaning (organic solvents), and vacuum cleaning (mechanical), only the first two will be discussed here. Little published information on vacuum cleaning is available. Kissa reports using a simulated procedure for removing excess particulate soils in sample preparation [24], and Robinson [72] discusses some technological aspects of vacuum cleaning of carpets.

The major importance of oils and fatty materials in soiling itself and in causing other soils to accumulate [11] suggests that the most effective cleaning procedure would involve using organic solvents to dissolve these important soil ingredients. Obviously problems of flammability, toxicity, and reuse of solvents make their application in home use intolerable at present. Dry cleaning has developed only as a commercial operation. All three of the problems with solvents are eliminated in the aqueous procedure, laundering, which is suitable for home use as well as for commercial operation. The main thrust of scientific inquiry has been towards optimizing soil removal by laundering, so this approach predominates in this review.

B. Laundering

1. Detergent Type

The most important elements in the multicomponent compositions referred to as "detergents" are the surfactants. These are amphipathic organic molecules generally consisting of a nonpolar, long-chain hydrocarbon segments attached to polar sites. Commonly used surfactants are classified according to the structure of the polar group at anionic, cationic, and nonionic [73]. Anionic surfactants include alkali metal salts of carboxylic acids (soaps), sulfonic acids, and alkyl sulfate esters; cationics are usually quaternary ammonium salts; nonionics include polyoxyethylene condensates, sucrose esters [74] and alkyl amine oxides [75]. The nonpolar segments of surfactants include straight-chain and branched alkyl groups (usually with 12 or more carbon atoms) or alkyl-aromatic groups.

Two major characteristics of surfactants which support their functions in detergent mixtures are their tendency to adsorb at interfaces and their ability to form micelles [76]. These capabilities combine in an effective washing system to provide for wetting, displacement of oils, dirt removal, soil suspension, and soil solubilization. Since not all these effects are attained by every surfactant, development of detergents has required optimization of the surfactant structure and combining it with supplemental additives.

An important variable affecting detergent action by all surfactants is their concentration in washing solutions. Preston [77] suggested that this concentration should be at least the level at which micelles are formed - the critical micelle concentration (cmc). McBain and Woo [78] supported this view for oily soils, but Goethe [79] disagreed based on results obtained with particulate materials. In a review of cmc effects, Harris [80] reported on a study by Demchenko which concluded that soil removal was initiated at detergent levels higher than the cmc. It has been suggested [81] that practical problems with soil removal from fabrics which are difficult to clean frequently arise from addition of too little detergent.

In a review of developments in detergent formulation, Jones [7] reported on variables affecting efficiency of anionic surfactants in soil removal. Some of the effects noted include:

Straight alkyl chains on benzene sulfonates are superior to branched chains

Effective detergency of benzene sulfonates begins with an alkyl chain length of 10 carbon atoms and improves to a maximum at 14-16 atoms

Surfactant effectiveness differs with soil type

p-Alkylbenzene sulfonates are superior to o-alkyl compounds

Straight chain carboxylates and sulfonates have similar detergent activity

Adsorption of surfactants on both soil and fibers affects detergent action

Soil dispersing power is a necessary but not sufficient surfactant characteristic for good detergency.

Cationic surfactants have the unique possibility of providing germicidal effects along with their cleaning action. This makes them useful in applications where antiseptic conditions must be maintained. Since the surfaces of most fibers and particulate soils are negatively charged, levels of cationic surfactants in excess of those needed to neutralize these charges would be expected to be needed before efficient detergent action was obtained [82].

H. T. PATTERSON AND T. H. GRINDSTAFF

In direct soil removal comparisons of anionic and cationic surfactants Grindstaff et al [16] showed that 85% of the hydrophobic carbon soil applied to a cellulosic surface was removed by 0.01M cetyltrimethylammonium bromide under washing conditions that produced 80% removal by 0.01M sodium lauryl sulfate. Both surfactants removed about 95% of the glyceryl tristearate applied to cellulose film, but the cationic compound removed 82% of this fatty soil from nylon, whereas the anionic product removed only 28%. It is probable that excess amounts of both surfactants were used in these studies.

Recent reviews by Harris [80] and Schönfeldt [83] have summarized extensive studies of the application of nonionic surfactants in detergent formulas. Some of the important findings of these investigations include:

Optimum soil removal activity for nonionic surfactants produced by condensing ethylene oxide with a normal, straight-chain aliphatic hydroxy compound is obtained with 12-14 carbon atoms in the chain and about 10 ethylene oxide units

Branched alkyl chains give less efficient detergent action than straight chains

Aromatic derivatives such as octyl- and dodecylphenol provide effective nonionic surfactants when condensed with about 10 mols of ethylene oxide

Nonionic surfactants soluble in dilute solution at room temperatures can become insoluble at a higher temperature (cloud point), and their detergent action is optimized close to this condition

Low levels of nonionic surfactants form micelles in water (believed by many to be a necessary requirement for good detergency) so that the amount of these compounds needed for optimum soil removal is less than that of anionic and cationic amphipaths

Effective soil solubilization shown by nonionic surfactants is an additional removal mechanism not available with ionic materials

Nonionic and anionic surfactants combined in detergent mixtures can give more effective soil removal than either surfactant alone

Many of the materials used as "builders" with anionic surfactants are also advantageous with nonionics and provide the additional benefit of permitting formulation of a solid, powdered detergent with liquid surfactants.

Laboratory comparison of soil-removal capabilities of surfactants from the three major classes by Fort et al [15] provided the results shown in Table 4. These data show an overall advantage for the nonionic surfactant over the ionic products. With the hydrophilic cellulose film almost all the soil was removed by all three surfactants, but for the more hydrophobic surfaces this was true only for the nonionic material. Even the fluoropolymer film was cleaned. These data

TABLE 4

Soil Removal from Polymer Surfaces[a]

Surfactant	% Glyceryl tristearate removal[b] from[c]:			
	Cellulose	Polyethylene terephthalate	Polyamide	Tetrafluoro-ethylene
Sodium lauryl sulfate	92	3	28	22
Cetyltrimethylammonium bromide	93	7	82	26
Nonylphenylpolyoxyethylene	94	99	99	96
Sodium palmitate	--	78	--	--

[a]From [15].
[b]C tagged ester removed by 0.01M surfactants in 15 min at 60°C.
[c]Film substrates.

indicate a probable difference in removal mechanisms among the surfactants.

A reported comparison [16] of the relative effectiveness of commercial anionic and nonionic detergent formulations in removing strongly adhering carbon black soil from fabrics is summarized in Table 5. Neither formulation removed significant amounts of carbon black from cotton fabric, but there was an advantage in the use of the nonionic material when synthetic sebum was mixed with the solid soil applied to a polyester staple fabric. Both detergents were similarly most efficient in cleaning a polyester continuous filament fabric. The potential soil solubilization ability of nonionic surfactants was of no benefit with particulate soils except when synthetic sebum was also present. However, these data indicate that a particulate carbon soil can be removed at least as well by a nonionic as by an anionic detergent mixture.

2. Builders

Nonsurfactant additives known as "builders" have long been used to improve detergent action of surfactants [83]. Some of the materials used include:

Alkali salts (carbonates, silicates, phosphates and borates)

Alkali hydroxides

Inorganic salts (sodium sulfate, etc.)

TABLE 5

Particulate Soil Removal From Fabrics[a]

	% Carbon black removed[b]					
	Cotton fabric		Polyester fabrics			
			Staple		Continuous filament	
Detergent solution (0.33%)	No sebum[c]	Sebum present	No sebum	Sebum present	No sebum	Sebum present
Built commercial anionic	9	11	25	8	66	67
Built commercial nonanionic	0	0	33	30	66	63

[a]From [16].
[b]15 min wash at 60°C.
[c]Synthetic mixture - 30% stearic acid, 30% glyceryl tristearate, 20% octadecanol and 20% octadecane.

Hydrophilic polymers (sodium carboxymethyl cellulose, polyvinyl alcohol, polyvinyl pyrrolidone)

Metal ion sequestering agents (ethylenediamine tetraacetate, nitrilotriacetate)

In his review of builder action, Jones [7] emphasized that no clear-cut explanation of builder action has been developed. This is probably because of the multiplicity of effects involved. Benefits from the addition of alkaline products have been confirmed in experiments with cotton fibers [84] and fabrics [85]. Among the possible explanations for this effect is that fatty acids in soils may be converted to soaps and then participate in further soil removal. Addition of neutral salts is known to reduce the critical micelle concentration of surfactant solutions [86], which could make them more effective as detergents.

Scott [87] confirmed this possibility in studies showing that adding electrolytes to anionic surfactants improved their detergent action on fatty soils depending on their degree of ionization, the pH of the solution, and the valence of the added cation. Lange [88] showed that salt addition at low levels (0.01N) increased detergent action most for low surfactant concentrations (i.e. 1×10^{-3} mols/liter) of sodium dodecyl sulfate and not at all for higher levels (1×10^{-2} mols/liter). He also showed that the effect was apparent with short aliphatic chains (10-12 carbons) but not for longer chains (14-16 carbons). At salt concentrations greater than about 0.03N, detergent action of anionic surfactants was reduced.

The importance of hydrophilic polymers to builder activity will be considered below under soil redeposition (Sec. VII B). Removal of metal ions by sequestering agents can prevent precipitation of carboxylate soaps and some straight-chain sulfonates in hard waters [7].

3. Temperature

As in all chemical and physical processes, temperature has an important influence on detergency. Raising temperature during laundering may be generally beneficial, but the reverse effect has occasionally been observed. Factors important to soil removal dependent on temperature include:

Soil diffusion (see Sec. V B-3)

Fabric penetration by fatty soils (melting and viscosity reduction)

Wetting and sorption at interfaces

Softening and swelling of textile surfaces

Solubility and micelle formation of surfactants.

Evidence supplied by Scott [87] showed that, with a built anionic surfactant and cotton fabric, removal of low-polarity solids such as octadecane and tripalmitin suddenly increased from a slow to a more rapid rate as the wash temperature was increased through their melting points. More polar solids, octadecanol and stearic acid, showed a continuously increasing rate of removal with increased temperature, starting well below their melting points. The difference in behavior was attributed to interactions between surfactants and solid soils to produce mesomorphic phases which could be swept from the surface by the flow of detergent solution. In the same study, Scott demonstrated that a nonionic surfactant could penetrate the surface of solid octadecanol to form mesomorphic phases at a temperature 10° below its melting point and thus improve its removal rate.

Additional observations by Powe [89] showed that removal of liquid mineral oils from cotton by a built anionic detergent increased by only about 5% as the temperature increased from 80 to 180°F. Removal of a solid petrolatum melting at 108°F increased by about 13% in the same range and a paraffin melting at 118°F increased from 3% at 80° to about 40% at 130°. In the latter case the rapid rise in removal started well below the melting point and continued above it.

Studies by Kissa [19] showed that the logarithm of the amount of used motor oil remaining on a fabric after laundering decreased, like the logarithm of the viscosity of soils or emulsions, with the decreasing reciprocal of the absolute temperature. The similarity of the two relationships suggested to him that increased washing temperature facilitated oily soil release by decreasing viscosity of the soil or its aqueous emulsion.

The studies of Berch et al [51] reported above (Sec. V B-3) demonstrated increased interaction of some soils (i. e. oleic acid) with surface coatings at an elevated temperature. This was established for an acrylic film, but not for silicone or perfluoroacrylate films. Oily soils which become strongly attached to coatings at elevated temperatures would be difficult to remove by laundering.

Increased water temperature can increase swelling and softening of the surfaces of textiles being washed [90]. Expansion of the surface may lead to dislodgment of some soils [87] but at the same time the surfaces can become tacky leading to stronger soil-fiber bonding. An additional consideration is that swollen fibers could reduce the available capillary space for flushing action by the detergent solution, particularly when dealing with tight fabrics.

Temperature effects on the colloidal characteristics of surfactants in water are considerably different for ionic and nonionic materials. The critical micelle concentration (cmc) of sodium dodecyl sulfate increased by 40% between 35 and 65°C [79] and that of potassium myristate by 27%. In contrast, for n-decanol condensed with six mols of ethylene oxide, the cmc dropped by 44% as the temperature increased from 15 to 45°C [91]. Increased temperature decreased the micellar size of ionic compounds slightly, but for nonionic materials the reverse effect was observed. For example, with n-hexadecanol condensed with nine mols of ethylene oxide the increase was about five-fold for a temperature rise of 25 to 53.5°C.

4. Mechanical Energy Input

Consideration of the effect of mechanical energy input on soil removal from textiles involves the intensity of agitation and the time of the washing cycle. An attempt was made by Bacon and Smith [92] to equate the removal of an oily particulate soil from cotton fabric with the important washing variables including both time of washing and agitation intensity. For a particular fabric at a fixed temperature and specific detergent their equation reduced to the form:

$$S = K(CFT)^n$$

where

S = % soil removal determined by reflection

K, n = constants for specific detergent

C = detergent concentration

F = energy of agitation

T = washing time

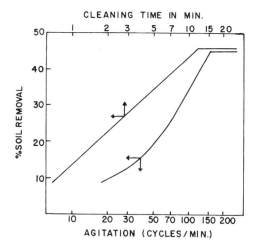

FIG. 9. Effect of agitation and cleaning time on soil removal. Data from [93].

Agitation was varied by changing the rate of rotation and the number and mass of balls used in a tumbling type of washer (launderometer). Plotting soil removal against time or agitation energy at constant detergent concentration produced a straight line with slope and intercepts controlled by values of K and n for the detergent.

Data from Linfield et al [93] may be replotted as shown in Figure 9 to indicate a linear relationship between soil removal and the logarithm of washing time rather than the relationship predicted by the Bacon and Smith equation. Dependence of soil removal on the logarithm of agitation deviated somewhat from linearity, but reached a limiting removal value as did the washing time results.

Kissa's results [19] demonstrated that the relationship between soil removal and the logarithm of washing time generally involved three linear steps with different slopes: an induction time, a rapid soil removal period, and a final period with essentially constant soil retention. He also found that oily soil removal from hydrophobic fibers was more readily improved by increased agitation when the fibers were smooth than when they were rough. A further conclusion from this study was that mechanical work causing hydrodynamic shear and flexing can compensate for slow spontaneous soil release from smooth hydrophobic fibers.

Working with wool fabrics which shrank because of felting during washing, Brooks and McPhee [94] showed a difference in the response of removal to agitation intensity depending on the type of soil involved. Oily soil was more easily removed and less dependent on agitation level

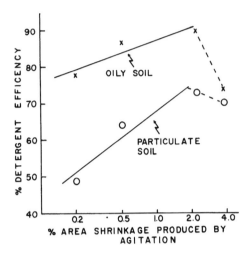

FIG. 10. Effect of agitation and soil type on detergent efficiency.
Data from [94].

than particulate soil. Since felting or shrinking increased with increasing agitation, shrinkage level could be used as an index of agitation to
provide the plots shown in Figure 10. Removal increased with agitation for both soil types to a maximum level and then declined as the
fabric tightness imposed by shrinkage achieved a level at which the
washing solution no longer moved freely in and out of fabric capillaries.

C. Dry Cleaning

Dry cleaning is an important commercial procedure for textile
cleaning. It has been improved to make possible individual coin-op-
erated machines, but it has not yet been developed to make widespread
household use possible.

Solvent selection cannot be made solely on the basis of soil removal. W. J. Stoddard and L. E. Jackson [95] drew up requirements
for a petroleum fraction (Stoddard solvent) which included specifications for clarity, color, odor, flash point, corrosion, distillation
range, nonvolatiles, acidity, mercaptan level, and degree of unsaturation. Although carbon tetrachloride was once used, its high toxicity and corrosiveness led to its replacement by perchloroethylene.
More recently [96] trichlorotrifluoroethane (known also as solvent
112, Valclene, or Arklone) and trichlorofluoromethane (solvent 11)
have been introduced as dry cleaning solvents.

Some components of textile soils which are easily removed by water because of their solubility, such as sugars and salts, are not removable by solvents for fats. It is necessary, therefore, to add small amounts of water with surfactants to keep it solublized in order to remove water soluble soil components. Martin and Fulton [95] reported that adding up to 3% of a detergent to a dry cleaning solvent did not increase the amount of salt removal. However, increasing the water content from 0.17 to 0.29% with 3% detergent present increased salt removal from 57.1 to 71.5%.

In his review of recent dry cleaning progress, Perdue [96] emphasized the importance of automatic adjustment of water level during dry cleaning to accommodate the moisture regain of the fibers involved. Sufficient water should be added to improve cleaning, but not enough to cause shrinkage, loss of creases or felting of wool fabrics. The partial pressure of water vapor in equilibrium with solvent surfactants and fabrics is expressed in a ratio to the saturation pressure termed "solvent relative humidity" (SRH). A value of 75% for SRH gives good removal of water-soluble materials. In a representative formulation, about 0.5 to 2.0% water (based on fabric) and about 0.1% detergent are used.

In considering the possible advantages of the fluorochemical solvents, Perdue mentioned their relatively low tendency to remove dyes and alter fiber properties, which makes them more like petroleum distillates than like perchloroethylene. They have low heats of evaporation and low boiling points so that operations must be in closed systems to prevent losses. The toxicity of solvent 113 is low enough to permit 1000 ppm vapor concentration rather than the 100 ppm permitted for perchloroethylene.

As with aqueous soil removal, dry cleaning requires addition of surfactants to maintain suspension of removed insoluble soil particles. Selection of optimum detergents has been studied by Csuros et al [97], Smith et al [98], and Meek [99].

VII. SOIL REMOVAL

A. Soil Release

References are made throughout this review to a variety of mechanisms postulated to be important to soil removal [5, 101, 114]. At the risk of some repetition, these will be included with additional hypotheses not yet discussed to provide a more comprehensive view. The most common source of confusion in the evaluation of soil-removal results is the inclination to assume that only one mechanism is needed to explain them. The much-discussed complexity of this operation necessitates consideration of the simultaneous contributions of several operations to the overall results. These may include:

"Rolling up" of liquid soils

Soil penetration followed by solublization and/or emulsification

Dislodgment of solid soil chips

Fiber-surface swelling leading to increased mechanical shear

Chemical reaction to produce more easily removed compounds

Modification of fiber and soil surface charges

Mechanical shear

Soil-mixture effects.

The concept of "rolling up" of oily soils first advanced by N. K. Adam [100] has been thoroughly reviewed by Schott [101]. By this mechanism, an oily film on a fiber immersed in a dilute surfactant is gradually displaced by adsorption of the surfactant on the fiber surface. The contact angle of the oil at the conjunction of fiber, oil and surfactant solution increases from a low value (close to zero) to a value approaching 180° which causes the oil to roll up into a drop. Such drops are easily dislodged by mechanical forces. Examples of this type of mechanism have been amply documented for special conditions by photomicrography [17, 84, 100]. With this mechanism, it is obviously necessary that soils be liquid during removal and that fiber surfaces be hydrophilic to promote surfactant adsorption. This latter requirement has encouraged development of many coatings and processes to increase fiber-surface hydrophilicity [102, 103] which in turn were found to promote soil accumulation because of their higher surface energy [104, 105]. As mentioned above (Sec. V A) the dual action hydrophobic/hydrophilic surface developed by Sherman et al [56] represents a useful compromise answer to this difficulty.

An alternative to soil removal by attack at the fiber-soil interface is to penetrate the external surface of the soil. Activity of wash solutions at this surface is limited only by their ability to permeate the interstices of soiled textiles. This mechanism augments the importance to soil removal of fabric geometry and fiber changes induced by their exposure to washing liquids such as increased bulk or surface swelling. Penetration of soil layers is favored by solubility of the wash solution surfactants in soils which gives an advantage to nonionic materials [106] over ionic surfactants. However, even with the latter compounds some degree of interaction with polar soil components can be achieved to produce viscous complexes which can be removed from the surface by mechanical shear [107]. Nonpolar soil components may be incorporated into surfactant micelles to produce "nonpolar solubilization" and eventually emulsification of the soil. This effect requires that surfactants be present at least at concentrations high

TABLE 6

Effect of Substrate on Soil Removal[a]

. Film substrate	% Soil removed[b]	
	Sodium lauryl sulfate	Nonylphenylpoly-ethylene glycol
Cellulose	97	95
Nylon	29	99
Polyethylene terephthalate	6	99

[a]From [15].

[b]15 minutes in water at 60°C, glyceryl tristearate soil, 0.01 M surfactant.

enough to form micelles (i.e., equal to or greater than the critical micelle concentration.

Direct evidence that soil penetration can be a major factor controlling soil removal is provided in a comparative study by Fort et al [15] using radio-tagged soils and films as shown in Table 6.

For the unbuilt anionic surfactant, sodium lauryl sulfate, dependence of detergency on the hydrophilic character of the surface was so dramatic that the removal might be logically interpreted by a rolling-up mechanism. With the unbuilt nonionic surfactant, there was no dependence of removal on the substrate so that a soil penetration removal mechanism seems more probable.

Soil removal by dislodgment of solid chips or flakes becomes important to washing at low temperature. If the soil is present as a thin, relatively nonextensible, solid film during washing, expansion of the substrate can cause dislodgment of chips of material. The plausibility of this mechanism has been demonstrated with radio-tagged soils by filtering wash liquids to show that increased levels of filterable solids are present when the soiled substrates are swollen by water [87]. Soil removal results supporting this hypothesis [15] are summarized in Table 7.

The considerable swelling possible with cellulose promoted removal of a large amount of soil by both surfactants. Essentially no soil was removed by the unbuilt anionic surfactant from the nylon and polyester surfaces which are less sensitive to water. The substantial amount of soil removed by the nonionic surfactant from surfaces with low moisture regain is explained by penetration of the solid soil, as already discussed. In the removal of tripalmitin from cotton fibers at temperatures below its melting point using a potassium oleate solution, Scott [87] proposed that the detergent solution entered

TABLE 7

Relationship of Moisture Regain to Solid Soil Removal

| | % Soil removed[a] | | |
Substrate	Sodium lauryl sulfate	Nonylphenyl-polyethylene glycol	Moisture regain (%)
Cellulose	73	71	12
Nylon	3	31	4.5
Polyethylene terephthalate	1	43	0.5

[a]15 min, 20°C, glyceryl tristearate soil.

cracks and dislocations in the soil crystalline aggregates and aided in breaking up the solid soil on fiber surfaces.

Swelling of surface coatings [90] has been postulated to contribute to soil removal even at temperatures high enough to melt fatty materials. This has been ascribed to the introduction of a mechanical-shear effect which disrupts the soil-surface interaction.

Possibilities for chemical reactions during washing may contribute to soil removal under the proper conditions. Free fatty acids in soils can be converted to alkali soaps and assist in removal of their soil components [108]. At sufficiently high temperatures and alkalinity, fatty esters may be hydrolyzed to produce soaps, although this would probably be important only in laundry procedures involving boiling at high pH levels. Oxidation of soils by bleaches can destroy colored components and also break down large molecules into smaller segments with polar sites. Even particulate carbon can be oxidized by bleaching to a more hydrophilic and readily removed form [16]. Enzymes have been used to split large and frequently colored soil molecules into less colored and more readily removed fragments [109]. Tomiyama and Iimori [110] reported a specific interaction of built anionic detergents with a protein component of natural soils which greatly improved their effectiveness. The usefulness of chemical reactions in assisting soil removal thus appears related to the properties of specific soil components.

Electrical charge effects on soil removal affect particularly the separation of particulate materials from fiber surfaces. In his review of washing theory Kling [82] proposed that such particles were "snapped in" or held in place on fibers by a minimum in the potential energy-distance from the fiber curve produced by the combination of the van der Waals-London attraction forces and the electrical forces of repulsion. He cites references to support that both fibers and soil particles have negative charges in water and these charges are strengthened by hydroxyl ions, multivalent anions, and anionic surfactants in washing solutions. The result of this strengthening of electrical repulsion facilitates particulate soil removal. To explain the functioning of nonionic surfactants, Kling reports the results of ultracentrifuge studies which showed that adhesion of soil-model particles to cellulose in water was markedly reduced by nonionic surfactants, although not as much as by anionic materials. Schott [111] and Cutler et al [112] have also published extensive reviews in this area of study.

Mechanical shear is a consequence of the intensity of agitation during washing and has already been discussed. The droplets produced by rolling up, the soil-surfactant cómplexes produced by penetration and solid soil particles must all be swept away from fiber surfaces to produce good detergency. The velocity gradient of a liquid flowing past a solid surface decreases essentially to zero as the surface is approached more closely [112]. This means that despite increases in agitation, the shear force acting on small solid particles is very low when they do not project much above the surface of the fiber.

The broad range in the ease of removability of individual soil components leads to some advantages when they are present as mixtures in natural soils. A relatively easily removable, fatty soil component enhances the release of more difficultly removable materials in a mixture [15]. Fatty soils may act as cements to bind particulate solids to textiles, and since they are relatively easier to remove, they can release the solids for suspension in washing fluids [11]. Even though precise control of the situation is not possible in most instances, rubbing of soil stains with an additional soil such as vegetable shortening or a concentrated liquid detergent to aid in their removal is a common practice.

B. Soil Redeposition

A widely recognized difficulty with efficient soil removal is that after initial separation, soils suspended in wash liquids may redeposit on fibers before they are completely removed from the textile assembly. Sorption of detergent components at fiber and soil surfaces provides barriers [79] to readherence of the soils when they are present at a sufficiently high concentration. For oily soils, these adsorbed layers increase the contact angle of droplets to close to 180° [50] so that they have essentially no tendency to rewet the fibers on recontact. With charged ionic surfactant molecules, the adsorbed layers on fibers and soil particles can produce an electrical repulsion which opposes redeposition. In his review of washing theory, Kling [82] emphasizes

that phosphate builders can supplement the suspending action of surfactants for soils by adsorbing on hydrophilic clay particles and keep-
ing them suspended.

In a study of soil deposition and transfer using cellulose and polyester films, Grindstaff et al [113] showed that at concentrations above
the critical micelle concentration (cmc), an unbuilt nonionic surfactant
inhibited deposition of an oily soil on a polyester surface and reduced
its transfer from cellulose to polyester film. At concentrations below
the cmc, it had little effect. An unbuilt anionic surfactant was only
moderately effective in preventing deposition above its cmc, and below this concentration it actually increased the amount deposited.
These results suggest that wash solutions containing suspended oily
soils should be removed as completely as possible from fabrics before initiating rinsing.

In a review of soil redeposition studies, Davis [114] mentioned
the research of several workers which emphasized the fact that variables influencing this process are affected by the manner in which soils
are applied. Hensley [115] deduced that transfer of soils from soiled
to clean cotton fabrics in a laundry bath produced results more closely
related to practical experience than pick-up of soils by fabrics from
their aqueous dispersions. Evaluation of the effectiveness of a series
of hydrophilic polymers in preventing soil deposition showed that polyvinyl alcohol, polyethylene glycol, polyvinyl pyrrolidone and carboxymethyl cellulose all reduced soil pick-up from aqueous dispersions.
Only carboxymethyl cellulose was effective in preventing soil transfer in a washing bath. He proposed that the size and agglomeration
of soil particles differed in the two processes. Smith et al [98] reported similar experiences in the evaluation of detergent effects in
dry cleaning. Their data indicated that the most effective detergents
for soil removal were also those which showed the least transfer from
soiled to clean fabrics in the same bath.

Reduction of soil redeposition by sodium carboxymethyl cellulose
has been extensively examined as indicated by Evans and Evans [116].
Their comprehensive research using a radio-tagged polymer indicated
that strong adsorption of the polymer by the fibers rather than by the
soils was necessary to prevent sorption of carbon black particles. The
necessary level of adsorption for good antiredeposition by strongly hydrophilic cotton fabrics exceeded by far that of nylon fibers. In two
additional papers by Johnson and coworkers [117, 118], the general
colloid stability theory of Derjaguin, Landau, Verwey and Overbeek
[119] was applied to this system. The combination of repulsive elec-
trostatic and attractive van der Waals forces was found adequate to
explain the antiredeposition activity of sodium carboxymethyl cellulose based on electrophoresis and sedimentation results.

These studies revealed that soil characteristics such as particle
size and surface charge were particularly important in obtaining

maximum reductions in redeposition. Since such variations are uncontrollable in practical situations, the results of carefully controlled experiments are only broadly applicable to everyday laundry problems. Pepperman et al [120] improved the launderability of resin-finished cotton fabrics by incorporating sodium carboxymethyl cellulose in the coatings. Soil redeposition reduction with polyesters has also been approached by increasing surface hydrophilicity. Perry [121] reports that a polyether-polyester copolymer may be diffused into the surface of polyester fibers to provide durable soil release and antiredeposition action.

C. Soil Hiding

By manipulating light reflection to give the appearance of cleanliness, the presence of soils on textiles may be covered up without removing them. One approach to this goal is through alteration of the reflective characteristics of the original fibers. East and Ferguson [122] found that changing cellulose acetate fibers from round to trilobal cross sections improved appearance in the presence of soils because of increased scattering of light reflected from their surfaces. The effect can be augmented as shown in further studies [123] by adding delusterants to the fibers internally or on their surfaces. The presence of oils in applied soils can counteract this effect when the oil rather than the fiber is the reflecting surface. In this way the refractive index differential with air is reduced.

Another procedure for modifying the optical properties of soiled textiles to give the appearance of cleanliness is to incorporate materials which fluoresce in the ultraviolet portion of the spectrum [124]. Such materials transform incident ultraviolet light partly into visible white light with a bluish tinge. These brightening agents may be sorbed from suspensions in the detergent wash bath [125, 126] or blended with molten polymers before spinning [127]. Brighteners for use in detergents include substituted pyrazolines, coumarins, stilbenenaphthotriazoles, diimidazolyl-ethylenes and dibenzoxazolylthiophenes. Materials stable enough to resist the high temperatures of spinning nylon or polyesters include substituted oxadiazolyl-stilbenes, triazinyl-stilbenes and naphthotriazolylphenylcoumarins. All of these compounds include groups which provide many opportunities for electron resonance. This optical approach can obviously only provide effective improvement in appearance with low levels of transparent soils.

VIII. IMPLICATIONS FOR FURTHER STUDY

Removal of soils from textile surfaces involves application of the principles of surface chemistry, adhesion, mechanics, and hydraulics to the separation of heterogeneous combinations of particulate solid and oily liquid materials from complex fibrous substrates. Understanding of the multiplicity of simultaneous interactions involved is a continuing challenge for further investigation. Some effort has been made to apply statistical methods in this field [128], but a multivariable attack in depth on this problem using modern computer techniques is needed to elucidate the many interactions not yet recognized or understood.

This review of previous work suggests several concepts, most of them already considered in part, but which merit particular attention as further studies are undertaken. These include:

1. Soil removal during laundering and dry cleaning probably takes place by several mechanisms acting simultaneously.

2. The problems of interpreting and applying results from experiments with synthetic soil mixtures arise from selection of the ingredients and the method for their application. Some suggestions for overcoming these difficulties are:
  a. Multiple soiling-cleaning cycles are needed for practical evaluations of changes in any of the operating variables. Since none of the present cleaning procedures removes all the applied soil, the significant substrate is not a clean fabric, but one contaminated with soil and cleaner residues.
  b. Synthetic soils should be applied at levels close to those obtained with normal exposure to natural soils.
  c. Soiled samples need aging before cleaning to complete soil-fiber interactions.
  d. Selection of particulate materials for mixing with oily soils should be made with the idea of duplicating the highly viscous pastes created by many natural dirt components and not necessarily obtained with clay particles frequently used in synthetic mixtures.
  e. A detailed assessment of the location and condition of soils remaining after cleaning would be useful. Residues may be continuous or discontinuous surface layers, particles lodged in fiber and/or fabric interstices, oils diffused into fibers, or a combination of all of these.

3. Consideration should be given to the changes in physical as well as chemical properties of textile fibers under cleaning conditions. Fiber swelling and changes in modulus may create geometric alterations in fabric structure which can be particularly important in fiber blends.

4. Advances in liquid and gas chromatographic techniques offer potentially useful applications to the study of soil removal.

5. Recent studies by Zettlemoyer and coworkers [129, 130, 131] explore the criteria affecting stability of the contact line developed in a four-phase, oil-water-solid-vapor system. Since this combination of phases describes the situation encountered in detergency, implications of this multiphase interaction concept to soil removal should be examined fully.

IX. POSTSCRIPT

In the interval between the time covered by this review and its publication, several interesting articles pertinent to soil release by textile surfaces have appeared.

Rounds, Purchase, and Smith [132] used radio-tagged fatty soil constituents and scintillation counting to establish the adverse effect of clay soil particles on fabric yellowing by unsaturated oily soils.

An extensive review of surface and interfacial tension of polymers by Wu [133] covers methods of measurement, effects of polymer variations and theoretical interpretations of results from many sources.

A further study by Jacobasch with Schumann [134] examined the influence on soil removal of adsorption of nonionic and anionic surfactants on the soil and fibers. They found that, in contrast to results on rayon fibers, soil removal from polyester fibers was proportional to the adsorbability of the surfactants. H. Kraus [135, 136] demonstrated that prewashing fabrics with a cationic surfactant followed by use of an anionic surfactant produced cleaner fabrics and markedly reduced the total amount of (nonphosphate) detergents required.

Many recent publications have examined the effects of soiling and laundering on flame resistance of fire retardant fabrics. Adler [137] identified factors such as fabric type, storage conditions, finishing and laundering conditions that must be considered in the evaluation of flame retardancy when laundering is included. Goynes, et al [138] made a microscopical survey of the structure of flame-treated cotton fibers and found that if the retardant were deposited unevenly it was removed by repeated laundering. Tests by Bullock and Carter [139] and by Simpson and Campbell [140] indicated laundering had little effect on flame resistance. However, Carfagno and Pacheco [141] showed that flammability increased if hard-water detergent or soap residues were not removed.

ACKNOWLEDGMENTS

The authors wish to express their appreciation to E. I. du Pont de Nemours and Company for its support and assistance in obtaining the information used in this work and for permission to publish it. Helpful comments from E. Kissa are also appreciated.

REFERENCES

1. O. Oldenroth, Chemiefasern, 18, 286 (1968).
2. K. Linder, Fette, Seifen, Anstrichm., 65, 96 (1963).
3. C. B. Brown, Research, 1, 46 (1947).
4. C. A. Bowers and G. C. Chantry, Text. Res. J., 39, 1 (1969).

5.   W. C. Powe in "Detergency Theory and Test Methods, Part I, "
     (W. G. Cutler and R. C. Davis, eds.), Dekker, New York, 1972,
     pp. 33-34.
6.   H. L. Sanders and J. M. Lambert, J. Amer. Oil Chem. Soc.,
     27, 153 (1950).
7.   T. G. Jones in "Surface Activity and Detergency," (K. Durham,
     ed.), Macmillan, New York, 1958, pp. 72-118.
8.   G. A. Byrne, J. Soc. Dyers and Col, 88, 64 (1972).
9.   P. A. Florio and E. P. Mersereau, Text. Res. J., 25, 641
     (1955).
10. S. Shimauchi and H. Mizushima, Amer. Dyest. Rep., 57, 462
     (1968).
11. T. Fort, Jr., H. R. Billica, and C. K. Sloan, Text. Res. J.,
     36, 7 (1966).
12. E. Kissa, Text. Chem. Color., 5, 249 (1973).
13. H. T. Patterson, unpublished data, 1969.
14. A. R. Martin and G. P. Fulton, "Drycleaning Technology and
     Theory," Textile Book, New York, 1958, pp. 22-23.
15. T. Fort, Jr., H. R. Billica, and T. H. Grindstaff, Text. Res.
     J., 36, 99 (1966).
16. T. H. Grindstaff, H. T. Patterson, and H. R. Billica, Text.
     Res. J., 37, 564 (1967).
17. C. L. Martin and B. A. Wood, Amer. Dyest. Rep., 60, 38
     (1971).
18. W. H. Rees, J. Text. Inst. (Trans.), 53, T230 (1962).
19. E. Kissa, Text. Res. J., 41, 760 (1971).
20. H. Schott, Text. Res. J., 39, 296 (1969).
21. M. D. Butler and H. T. Patterson, unpublished data, 1970.
22. "Amer. Assn. Textile Chemists and Colorists Technical Man-
     ual," 1970 Ed., Res. Triangle Park, N. C., Test Method
     AATCC 123 - 1970 (1970).
23. L. Benisek, T. M. Brown, G. C. East, and J. P. Ferguson,
     J. Text. Inst., 64, 189 (1973).
24. E. Kissa, Text. Res. J., 41, 621 (1971).
25. W. H. Smith, M. Wentz and A. R. Martin, J. Amer. Oil
     Chem. Soc., 45, 83 (1968).
26. K. Durham in "Surface Activity and Detergency," (K. Durham,
     ed.), Macmillan, New York, 1958, p. 225.
27. J. Compton and W. J. Hart, Ind. Eng. Chem., 43, 1564 (1951).
28. J. W. Hensley, M. G. Kramer, R. D. Ring, and H. R. Suter,
     J. Amer. Oil Chem. Soc., 32, 138 (1955).
29. I. Ruch, F. D. Snell, and L. Osipow, Ind. Eng. Chem., 45,
     137 (1953).
30. O. C. Bacon and J. E. Smith, Ind. Eng. Chem., 40, 2361 (1948).
31. L. Benisek, Text. Res. J., 42, 490 (1972).
32. G. B. Aspelin, W. B. Prescott, and R. Y. Krammes, Text.
     Chem. Color., 1, 233 (1969).
33. L. Benisek and G. H. Crawshaw, Text. Res. J., 41, 415 (1971).

34. P. S. Stensby, Soap Chem. Spec., 43, (5) 84 (1967).
35. E. Kissa, Text. Res. J., 43, 86 (1973).
36. B. E. Gordon and W. T. Shebs, J. Amer. Oil Chem. Soc., 46, 537 (1969); B. E. Gordon and E. L. Bartin, idem., 45, 754 (1968); B. E. Gordon, W. T. Shebs and R. V. Bonnar, idem., 44, 711 (1967).
37. S. Richards, M. A. Morris and T. H. Arkley, Text. Res. J., 38, 105 (1968).
38. D. A. Neutzal, C. W. Stanley, and D. W. Rathburn, J. Amer. Oil Chem. Soc., 41, 678 (1964).
39. G. Wlodarski and E. Balcerzyk, J. Text. Inst., 61, 506 (1970).
40. C. B. Brown, S. H. Thompson, and G. Stewart, Text. Res. J., 38, 735 (1968).
41. J. J. Cramer in "Detergency: Theory and Test Methods, Part I," (W. G. Cutler and R. C. Davis, eds.) Dekker, New York, 1972.
42. W. A. Zisman, J. Paint Technology, 44, 41 (1972).
43. J. H. Sewell, Mod. Plast., 48, (6) 66 (1971).
44. S. J. Gregg, "The Surface Chemistry of Solids," Reinhold, New York, 1951, p. 233.
45. W. D. Harkins, J. Chem. Phys., 10, 268 (1942).
46. F. M. Fowkes in "Wetting, SCI Monograph No. 25," Gordon Breach, Science Publishers, New York, 1967, p. 3.
47. D. H. Kaelble, J. Adhes., 2, 66 (1970).
48. T. Fort, Jr. in "Contact Angle, Wettability and Adhesion," Advances in Chemistry Series 43 (R. F. Gould, ed.) American Chemical Society, Washington, D. C., 1964, p. 302.
49. V. R. C. Mayrhofer and P. J. Sell, Angew. Makromolekulare Chem., 20, 153 (1971).
50. J. C. Stewart and C. S. Whewell, Text. Res. J., 30, 903 (1960).
51. J. Berch, H. Peper, and G. L. Drake, Jr., Text. Res. J., 35, 252 (1965).
52. W. P. Utermohlen, M. E. Rayn and D. O. Young, Text. Res. J., 21, 510 (1951).
53. S. Smith and P. O. Sherman, Text. Chem. Color., 1, 105 (1969).
54. ibid., Text. Res. J., 39, 441 (1969).
55. R. K. S. Chan, Text. Res. J., 40, 1059 (1970).
56. P. O. Sherman, S. Smith, and B. Johannessen, Text. Res. J., 39, 449 (1969).
57. H. Lange, Kolloid Z., 156, 108 (1958).
58. M. V. Stockelberg, W. Kling, W. Benzel, and F. Wilke, Kolloid Z., 135, 67 (1954).
59. K. Kanamaru, Kolloid Z., 168, 115 (1960).
60. H. J. Jacobasch, Faserfor. u. Textiltech., 20, 191 (1969).
61. J. Compton and W. J. Hart, Text. Res. J., 23, 158 (1953).
62. W. J. Hart and J. Compton, Text. Res. J., 23, 418 (1953).
63. J. Compton and W. J. Hart, Text. Res. J., 24, 263 (1954).
64. R. Tsuzuki and N. Yabuuchi, Amer. Dyest. Rep., 57, 472 (1968).
65. T. H. Grindstaff, Text. Res. J., 39, 958 (1969).

66. V. W. Tripp, H. T. Moore, B. R. Porter, and M. L. Rollins, Text. Res. J., 28, 447 (1958).
67. H. Peper and J. Berch, Amer. Dyst. Rep., 54, 863 (1965).
68. F. T. Pierce, J. Text. Inst., 28, T45 (1937).
69. ibid., Text. Res. J., 17, 123 (1947).
70. A. S. Weatherburn and C. H. Baley, Text. Res. J., 27, 358 (1957).
71. C. B. Brown, S. H. Thompson, and G. Stewart, Text. Res. J., 38, 735 (1968).
72. G. Robinson, "Carpets," Trinity, London, 1972, p. 252.
73. H. M. Gadberry, in "Industrial Detergency," (W. W. Niven, Jr., ed.) Reinhold, New York, 1955, pp. 9-50.
74. A. M. Schwartz and C. A. Roder, J. Amer. Oil Chem. Soc., 42, 800 (1965).
75. H. F. Drew and R. E. Zimmer (to Procter and Gamble Co.) U. S. Pat. 3,001,945 (1961).
76. K. Durham in "Surface Activity and Detergency," (K. Durham, ed.) Macmillan, New York, 1958, p. 1.
77. W. C. Preston, J. Phys. and Colloid Chem., 52, 84 (1948).
78. J. W. McBain and T. Woo, J. Phys. Chem., 42, 1099 (1938).
79. E. K. Goette, J. Colloid Sci., 4, 459 (1949).
80. J. C. Harris in "Nonionic Surfactants" (M. J. Schick, ed.) Dekker, New York, 1967, p. 706.
81. R. T. Hunter, C. R. Kurgan, and H. L. Marder, J. Amer. Oil Chem. Soc., 44, 494 (1967).
82. W. Kling, Melliand Textilber., 46, 957 (1965).
83. N. Schonfeldt, "Surface Active Ethylene Oxide Adducts," Pergamon, New York, 1969, pp. 386-441.
84. R. E. Wagg and G. D. Fairchild, J. Text. Inst., 49, T455 (1958).
85. F. Fumikatsu and T. Imamura, J. Amer. Oil Chem. Soc., 47, 422 (1970).
86. J. L. Moilliet, B. Collie, and W. Black, "Surface Activity," Van Nostrand, Princeton, 1961, p. 61.
87. B. A. Scott, J. Appl. Chem., 13, 133 (1963).
88. H. Lange, Kolloid Z., 169, 124 (1960).
89. W. C. Powe, J. Amer. Oil Chem. Soc., 40, 290 (1963).
90. C. E. Warburton, Jr., and F. J. Parkhill, Text. Chem. and Col., 5, 113 (1973).
91. P. Becher in "Nonionic Surfactants" (M. J. Schick, ed.) Dekker, New York, 1968, pp. 478-515.
92. O. C. Bacon and J. E. Smith, Ind. Eng. Chem., 40, 2361 (1948).
93. W. M. Lindfield, E. Jungermann and J. C. Sherrill, J. Amer. Oil Chem. Soc., 39, 47 (1962).
94. J. H. Brooks and J. R. McPhee, Text. Res. J., 37, 371 (1967).
95. A. R. Martin and G. P. Fulton, "Drycleaning Technology and Theory," Textile Book Publishers, New York, 1958, p. 108.
96. G. R. Perdue, Textile Progress, 2, 31 (1970).

97. Z. Csuros, J. Bozzay, A. Konig, and J. Szoeoka, Kolor Ertesito,
     7, 350 (1964); [through J. Soc. Dyers Col., 82, 158 (1966)].
98. W. H. Smith, M. Wentz, and A. R. Martin, J. Amer. Oil Chem.
     Soc., 45, 83 (1968).
99. D. M. Meek, J. Text. Inst., 59, 58 (1968).
100. N. K. Adam, J. Soc. Dyers Colour., 53, 121 (1937).
101. H. Schott in "Detergency Theory and Test Methods, Part I,"
     (W. G. Cutter and R. C. Davis, eds.) Dekker, New York, 1972,
     pp. 105-152.
102. R. J. Harper, Jr., J. S. Bruno, G. A. Gautreaux, and
     M. J. Donoghue, Text. Chem. Color., 2, 99 (1970).
103. H. B. Goldstein, Text. Chem. Color., 3, 8 (1971).
104. H. E. Bille, A. Eckell, and G. A. Schmidt, Text. Chem. Color.,
     1, 600 (1969).
105. R. Kleber, Textil-Praxis, 24, 42 (1969).
106. T. Nakagawa in "Nonionic Surfactants" (M. J. Schick, ed.) Dekker,
     New York, 1967, pp. 558-603.
107. A. S. C. Lawrence in "Surface Activity and Detergency,"
     (K. Durham, ed.) Macmillan, New York, 1958, pp. 158-192.
108. L. G. Johnston in "Industrial Detergency," (W. W. Niven, Jr.,
     ed.) Reinhold, New York, 1955, pp. 51-72.
109. T. Cayle, J. Amer. Oil Chem. Soc., 46, 515 (1969).
110. S. Tomiyama and M. Iimori, J. Amer. Oil Chem. Soc., 46, 357
     (1969).
111. H. Schott in "Detergency Theory and Test Methods, Part I,"
     (W. G. Cutter and R. C. Davis, eds.) Dekker, New York
     (1972), pp. 152-235.
112. W. G. Cutter, R. C. Davis, and H. Lany, idem., pp. 65-103.
113. T. H. Grindstaff, H. T. Patterson, and H. R. Billica, Text. Res.
     J., 40, 35 (1970).
114. R. C. Davis in "Detergency Theory and Test Methods, Part I,"
     (W. G. Cutter and R. C. Davis, eds.) Dekker, New York (1972),
     pp. 269-321.
115. J. W. Hensley, J. Amer. Oil Chem. Soc., 42, 993 (1965).
116. P. G. Evans and W. P. Evans, J. Appl. Chem., 17, 275 (1967).
117. G. A. Johnson and K. E. Lewis, J. Appl. Chem., 17, 283 (1967).
118. G. A. Johnson and R. A. C. Bretland, J. Appl. Chem., 17, 288
     (1967).
119. E. J. Verwey and J. T. G. Overbeek, "Theory of the Stability of
     Lyophobic Colloids," Elsivier, Amsterdam, 1948.
120. A. B. Pepperman, Jr., G. L. Drake, Jr., and W. A. Reeves,
     Amer. Dyest. Rep., 60, 52 (1971).
121. E. M. Perry, Amer. Dyest. Rep., 57, 405 (1968).
122. G. C. East and J. P. Ferguson, J. Text. Inst., 60, 441
     (1969).
123. Ibid., 64, 273 (1973).
124. K. J. Nieuwenhuis, J. Amer. Oil Chem. Soc., 45, 37 (1968).
125. R. Anliker, H. Hefti, H. Kasperl, J. Amer. Oil Chem. Soc.,
     46, 75 (1969).

126. R. Anliker, H. Hefti, H. Kasperl, and B. Milicevic, J. Amer. Oil Chem. Soc., 46, 523 (1969).
127. C. Eckardt and H. Hefti, J. Soc. Dyers and Col., 87, 365 (1971).
128. J. P. Narcy and J. Renaud, J. Amer. Oil Chem. Soc., 49, 598 (1972).
129. A. C. Zettlemoyer, M. P. Aronson, and J. A. Lavelle, J. Colloid Interface Sci., 34, 545 (1970).
130. M. C. Wilkinson, M. P. Aronson, A. C. Zettlemoyer, J. Colloid Interface Sci., 37, 498 (1971).
131. M. P. Aronson, A. C. Zettlemoyer, and M. C. Wilkinson, J. Phys. Chem., 77, 318 (1973).
132. M. A. J. Rounds, M. E. Purchase and B. F. Smith, Text. Res. J., 43, 517 (1973).
133. S. Wu, J. Macromol. Sci. Rev. Macromal. Chem., 10, (1), 1 (1974).
134. H. J. Jacobasch and U. Schumann, Faserfor. U. Textiltech., 24, 248 (1973).
135. H. Kraus, Tenside Deterg., 12, 137 (1975).
136. Ibid., 12, 200 (1975).
137. A. Adler, Text. Chem. Color., 7, 76 (1975).
138. W. R. Goynes, E. K. Boylston, L. L. Muller, and B. J. Trask, Text. Res. J., 44, 197 (1974).
139. A. L. Bullock and M. E. Carter, Am. Dyest. Rep., 64, 21 (1975).
140. H. N. Simpson and M. A. Campbell, Am. Dyest. Rep., 64, 28 (1975).
141. P. P. Carfagno and J. J. Pacheco, Proc. Annu. Meet. Chem. Spec. Manuf. Assoc., 60, 35 (1974).

Chapter 13

STAIN AND WATER REPELLENCY OF TEXTILES

Bernard M. Lichstein
Patient Care Division
Johnson & Johnson
New Brunswick, New Jersey

I. INTRODUCTION

Textiles are made repellent to withstand the stresses of use and of the environment. Thus, rainwear is treated to withstand a downpour directed at its outer surface. A firehose is tightly woven to contain water under pressure against its inner surface. Upholstery fabrics are coated with repellents which prevent soiling by aqueous and oily stains and by dry soils. Progress in repellent finishes has followed logical paths: from occlusive coatings to those that permit the fabric to breathe; from fugitive finishes to those that withstand repeated launderings and dry cleanings; from finishes that retain upon the fabric those stains which have penetrated the repellent barrier to those which permit the stain to be removed by laundering or dry cleaning. Each advance has been built on the previous one; and all are subject to the whims, tastes, and price constraints of the commercial world.

Finished textiles are for the most part hydrophobic. Nylons, polyesters, and wools are inherently water repellent; and finishing processes usually convert even hydrophilic textiles into hydrophobic ones. The gross effect of both oil- and water-repellent finishes is one of changing a hydrophilic surface into a hydrophobic surface and lowering its surface free energy. A finish which repels an oily stain by lowering the surface energy at a fabric below that of such a stain, will also repel an aqueous stain. Most soils, even if water borne, contain fatty or oily components.

For these and many other reasons, the division between oil and water repellency is artificial and will be treated as such in this chapter. Soil-release finishes are also repellent. However, they impart sufficient hydrophilic character to the surface so that stains which have not been repelled, and have penetrated the repellent surface barrier, can be reached by laundering solutions. These finishes will not be discussed in this chapter.

II. MECHANISM OF SOILING

A. Fluid Stains and Capillarity

A fluid stain's ability to wet a textile and its component fibers can be described by the equation for capillary pressure [1, 2],

$$p = \frac{2\gamma \cos \theta_a}{r} \tag{1}$$

where γ is the surface tension of the stain, θ_a = the dynamic value of the advancing contact angle as it rises in the capillary or falls forward on a tilted surface, r is an average radius of the air-filled capillaries whose space is changing as the liquid first pulls the capillary walls together and then expands them on completely filling the capillary.

The advancing contact angle is that which includes the fluid and air between the solid surface and a tangent to the fluid at its intersection with the air and the solid surface (Fig. 1). In a static system, with a smooth, horizontal surface, the angle is equal at all points of intersection of air and surface. In a dynamic system, involving a tilted, smooth surface or one which is titled by virtue of its roughness, the angle uphill of the advancing contact angle is the receding contact angle, θ_r.

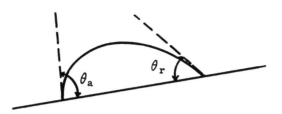

FIG. 1. Contact angles of a liquid on a tilted or rough, solid surface; θ_a = advancing angle; θ_r = receding angle.

In the capillary system of textiles, the advancing contact angle is a measure of the maximum value the fluid stain assumes on spreading or inflating. It is this angle that controls spreading and wicking. Stain repellency ratings are directly related to this value. The receding contact angle is a measure of the minimum value assumed by the drop as the front is moved toward the center of the drop by deflation or tilting. It predicts whether a drop will release fully on blotting or rolling or whether it will leave behind smaller, detached droplets. The greater the difference (hysteresis) between the two angles, the less likely that a drop will be able to roll freely despite increasingly high values of advancing contact angle [3-7].

Analysis of the variation of θ_a to water, with the surface finish of the fiber and the construction of the yarn, showed that a yarn is less susceptible to wetting if its surface energy is low and if the capillary spaces in the yarn are large [8]. Treatments, such as fluorochemicals, impart a low surface energy to the yarns which in turn results in a high value for θ_a (cos θ_a is low). Spontaneous wicking into the yarn is thereby prevented, and high pressures are needed to displace, with oil, the air between the yarns. Even higher pressures are needed to displace the air within the yarns [9-13].

When wicking can occur, Equation 1 predicts the capillary pressure will be greater for tight yarn constructions. However, the rate of wicking will be slower, as is predicted by Equation 2,

$$\frac{dl}{dt} = \frac{2 \gamma r \cos \theta_a}{8 \eta l} \tag{2}$$

where l is the distance traveled by the fluid, η is the viscosity, t is the time. Here, small values of r, for the small capillaries, and high values of η, for viscous oily stains, will slow stain penetration.

When staining does occur, it is at those points which present a combination of small capillary sizes, to cause high capillary pressures, and low interfacial area between fluid and air, to minimize the value of θ_a [14, 6]. These points are located in small surface irregularities of single fibers and at fiber cross-over points. The staining process is only possible if wicking can be initiated, that is when $\theta_a < 90°$. When $\theta_a > 90°$, with a good fluorocarbon finish, wicking doesn't occur easily and one need not be overly concerned with cleaning the stain immediately. The stain will not penetrate appreciably deeper into the fabric unless it undergoes some chemical change which changes its surface free energy or causes it to bond chemically with the fabric.

On laundering, lubricating oil releases more easily from continuous polyester filaments of circular cross section than from cotton [15]. Filament polyester or nylon fabrics have soil release capacities superior to those made of spun yarn [16]. Crimped polyester tends to soil more and release soil less readily than smooth polyester. Constructions which soil badly are those with frequent fiber cross-over points as occur in tight, high-twist, spun staple yarns.

Fluid stains do not spread uniformly even over smooth single fibers. They form alternating high and low spots which reflect the surface energies of fluid and fiber. However, if the fiber has wettable capillary systems along its axis, such as in crenulated rayon, or certain multilobal polyester and nylons, then fluids will wick spontaneously in these systems [2].

B. Dry Soiling

Dry, particulate soils also tend to build up in irregular voids and cross-over points [16]. The mechanism is mainly that of mechanical entrapment where soiling increases with the weight and complexity of the fabric. Complexity is related to surface irregularities, surface textures, fuzziness, and weave [17-19]. Simple fabrics, made of smooth fibers, soil less and by a sorptive mechanism [20]. It is not surprising that there is a direct relationship between the size of soiling particles and the size of the fiber irregularities which have become filled with the particles. In general, fine particles will soil fabrics to the greatest extent by mechanisms of entrapment and absorption.

Surface irregularities can in fact be filled, or presoiled, with preferred white or transparent pigments such as alumina, silica, or hard waxes. These prevent occupation of soilable sites by the more visible, darker soils [21]. The white pigments tend also to mask the presence of the darker soils by reflecting light multiply from internal surfaces of the irregularities.

Film-forming finishes tend to reduce particulate soiling by smoothing over surface irregularities. However, soft, rubbery finishes accumulate particles easily by adhesion and retain the soil by embeddment in the finish [21, 22]. Hard-finish films are resistant toward entrapment of particulate soils unless they present new surface irregularities by crazing and cracking [15].

Surface irregularities, and pits and crevices, need not always be smoothed over or filled in to increasing soiling resistance. Thus, when cottons were treated with perfluorodecanoic acid and its derivatives, microscopic examination showed the surface roughness had not changed from that before treatment. The resistance to soiling by particulate matter must have been due to reduction of the surface energy of the fibers and thereby the sorptive bonding of the particles [20].

C. Composite Soils

Retention of soils, which are composites of fluid and particles, such as dirty motor oil, lipstick, and soot, is predominantly due to the fluid component. Removal of the fluid, which acts as the carrier and cement for the particulate soil, results in easier cleaning of the fabric [24].

III. TEST METHODS

A. Oil Repellency Tests

Attempts to evaluate the repellency of finishes have followed the logical trend of more accurately reflecting performance to common stains under field conditions. It was demonstrated [25-28] that resistance to wetting by a cold water spray, or to penetration and spreading of drops of pure hydrocarbon oils, were not practical tests. Common, aqueous, and oily staining materials are complex mixtures which may stain even worse than their pure components. This was shown when salt, alcohol, surfactant, or a thickening agent was added to a cold solution of aqueous dye. Similar enhancement of staining was obtained when surfactant was added to suspensions of iron oxide or carbon black.

Nine staining materials were chosen (Table 1) as being representative of an original list of nineteen aqueous and eighteen oily materials. The more viscous materials were applied under controlled pressure. A shorter list of common household type staining materials was also developed where none were applied under pressure [12], since resistance to stains forced into the fabric is less easy to predict. However, despite this modification, tests with the actual staining materials are often discarded in favor of a series of hydrocarbon liquids of

TABLE 1

Representative, Common Aqueous and Oil Staining
Materials and Conditions of Application

Material	Pressure	Time
Hot coffee, 140°F	No	3 min
Ethanol, 25%, colored with 0.1% acid green 3, 70°F	No	3 min
Corn syrup colored with 0.1% acid blue 13, 70°F	Yes	30 sec
Beef gravy mixed with 25% Crisco at 140°F	No	3 min
French dressing, 70°F	No	3 min
Chocolate syrup, 70°F	Yes	30 sec
Corn oil colored with 0.1% solvent blue 36, 70°F	No	3 min
Crisco, 70°F	Yes	30 sec
Melted Crisco, 160-170°F	No	3 min

decreasing surface tension. The oil repellency rating of the fabric is determined by the liquid of lowest surface tension which will not wet and wick into the surface and therefore shows no penetration or spreading. The ratings reflect the surface treatment, the degree of surface coverage and the fabric geometry.

Three commonly used tests are the 3M test using Nujol [29], and the DuPont [12] and AATCC tests [30], both using a series of hydrocarbons. Drops of the test liquids are placed on the fabric for a designated time — 3 minutes in the 3M test and 30 seconds in the DuPont and AATCC tests. The repellency values given by the three tests give similar rankings in that a high repellency rating signifies resistance to wetting by test liquids of decreasingly low surface tensions.

Such lists now form part of the performance specifications of a finish. These specifications tell the consumer the minimum amount of finish (usually a fluorochemical) necessary for the projected end use; the rating obtained in water and oil repellency tests; the performance after laundering and dry cleaning. Typical performance specifications in terms of the AATCC test rating are given in Table 2. Note that minimum performance will be given by these lower limits of fluorocarbon add-on. Optimum performance will require higher add-ons.

TABLE 2

Typical Fluorochemical Manufacturer Specification Ratings (AATCC Minimum Rating)

Fabric	Spray Method No. 22-1964			Oil Repellency Method No. 118-1966G		
	Initial	After[a,c] Laundering	After[a,b] Dry Cleaning	Initial	After[a,c] Laundering	After[a,b] Dry Cleaning
Rain and outerwear						
Woven cotton and cotton/polyester blends	90	80	80	4	3	3
Woven fabrics of other fibers and their blends; knitwear and napped fabrics of all fibers; rubber or vinyl backed fabrics	80	70	70	4	3	3
Other apparel and home furnishings						
Apparel, drapery and slip-cover	80	70	70	4	3	3
Upholstery, necktie, ribbon, belting and rubber backed fabrics	80	--	--	4	--	--

[a] After 3 wash-wheel type launderings or after 5 home washings and 5 tumble dryings if "easy care" is claimed on the garment label.

[b] After 1 tumble jar or 3 Launder-O-Meter dry cleaning tests with perchloroethylene and with detergent.

[c] All tests may be omitted if the fabric is not intended to be cleaned by either method and no claim for these methods of cleaning is made.

Several quick tests and modifications have been developed [12] to determine the amount and adequacy (lack of pinholes) of finish coverage:

1. A blue dye is used in the hydrocarbon test liquids. This helps reveal wicking in light colored fabrics. No dye is used on dark colored fabrics.

2. A severe staining situation is mimicked by mechanically forcing blue-dyed 3-In-One household oil into the fabric. A fabric with a DuPont rating of 5 will usually withstand this test.

3. A "static stain test" helps determine the suitability of a particular fabric for a stain-repellent finish. Both water based and oil based stains are applied and then removed without forcing them into the fabric. Fabrics with different weaves, textures and piles will respond differently to this test even if the fluorocarbon add-on is the same. If the stains are wiped in, only the highest-level finishes can resist staining. Thus, fluorocarbon finishes are repellent but not to oily stains. Fabrics that respond with a rating of less than five, are also sponged with water to determine the stain release of the finish.

4. Ringing of the spot is sometimes encountered when dry-cleaning solvents are used to sponge or "spot" the fabric. This may be due to migration of the stain. When high-level fluorocarbon finishes are used it is more likely due to elements of the finish which are soluble in the dry-cleaning solvents. These may be dyes, waxes, softeners, and other adjuvants.

B. Water Repellency Tests

Approaches, similar to those outlined for oil repellency, have been applied to determine the ability of textiles to resist penetration and wicking of water [30] -- both before and after they have been subjected to conditions of projected use. In addition to water repellency, rainwear textiles have been tested for air and water-vapor permeability to relate to comfort factors [31-33].

Tests are use-oriented. They are arranged in two classes [27].

Class 1 consists of spray, sprinkling, and drop tests. These simulate effect of rain in wetting and penetrating the fabric:

1. The spray test detects whether or not a fabric has been treated so that it is resistant to wetting -- not to penetration.

2. The AATCC Slowinske raintester, the impact penetration tester, and the drop penetration tester all measure penetration of a simulated rainfall whose impact is related to the hydrostatic head acting on the spray. The first two simulate mild rainfalls. The drop penetration tester simulates the effect of a heavy rain striking the fabric in the same spot.

3. The Bundesmann water repellency test reflects service conditions by simulating rubbing and flexing on the inside of a garment during wear while it is rotated under a heavy downpour [34].

Class 2 consists of hydrostatic pressure tests. These are not necessarily applicable to water-repellent rainwear which resist wetting but permit penetration under a hydrostatic head. They are however useful for detecting pinholes in film-coated fabrics and are applicable to swellable types of fabrics and heavy, closely woven fabrics such as ducks, canvas, and firehose. The Mullen Burst tester has been adapted for this use.

Another test, widely used to measure the repellency of nonwovens, is the vented Mason Jar test [119]. The test square of fabric is held over the mouth of a one-quart jar between the gasket and a screwed-on metal ring. The jar contains a sufficient amount (about 600 ml) of 0.9% saline solution to give a 4.5 in. head. The vent hole at the bottom of the jar is sealed while filling. The jar is inverted and the vent opened. The time to wet through is observed.

IV. TEXTILE SURFACES

Proper fabric design concerns itself with reducing the free space between yarns and fibers to a size below that of a droplet of water or oil. Further concern is given to the requirements of spatial and penetration resistance to small droplets which are created by drops driven onto the fabric by the impact of rain or spray of liquid. In addition, it is wise to avoid pitfalls such as sewing with a hydrophilic, untreated thread or overcoating with another textile finish which will negate the repellent properties of the already repellent surface. Repellent linings are an aid in this direction.

A logical extension of fabric design, which resists gross penetration of water, is encountered in tightly woven tent material, water bags, and firehose. On absorbing water, the fibers expand to give an even tighter structure which resists leakage [35]. Related to reduction of porosity, is the observation that multiple layers of water-repellent fabrics resist water penetration much more than is predicted from the repellency of the individual layers [36]. They also afford greater energy adsorption on impact.

A review [37] of self-sealing fabrics describes the addition to the fibers of hydroxycellulose, which swells when wet, thus preventing passage of water [38]; the use of fine, immature cotton, which yields tighter packing [39]; and a special loom attachment to significantly increase the number of picks per inch [39].

Simulation of a tight weave, which prevents the passage of liquid water but permits the fabric to breathe, is obtained with microporous finishes. This has been done by milling a solid pigment into a synthetic rubber binder to give microscopic pores within the pigment [31, 32]. A pore-forming material can also be one that is soluble enough to be leached out of the film-forming polymer. Larger pores can be produced before leaching by heating the mixture to swell the admixed material [40]. Somewhat analogous to the use of swellable hydroxycellulose is the use of a water-swellable polyethylacrylate which is laid below the fabric surface in a discontinuous layer. The uncoated fabric surface retains its texture. On wetting, the polymer swells and forms a continuous-barrier layer [41].

Considerations which are secondary to providing a tight surface relate to the need to reduce surface roughness, since it has been demonstrated that bouth roughness as well as porosity decrease the contact angle of wetting [8]. Fuzziness and fraying in weaving is minimized by the appropriate degree of twist in the yarn.

The ability to shed water, in contrast to resistance to penetration by water, is best exhibited by rough, loose structures [42]. This is due to lessened capillary tension of the larger capillaries. Woolens, which are wet less easily, use the more open structure. Cotton and its blends use the tight structure.

Textiles with a given interfiber geometry, surface porosity, and roughness will respond differently to wetting if the histories of their exposure to the environment is different [43]. The presence of the residuals of finishing, dry cleaning, and laundering will modify the surface energy as well as the surface roughness and therefore the water repellency of the fabric. Soil, which is borne by water or oil, will also change the surface. Water, containing impurities or purposely added wetting agents, will have a lower surface tension. It will wet the surface of fabrics more easily [44].

The evaluation of processes for making textile surfaces water repellent reflects the search for permanence, ease of application, and low cost. All treatments require that the fabric be clean of natural impurities, sizes, alkalis, and soaps; thoroughly treated to leave no wettable, untreated surface; and free of other finishing agents which may mask or hinder the repellent effect.

V. TEXTILE FINISHES

A qualitative comparison of the efficacies and costs of repellent finishes is given in Table 3. Durable repellents are used mainly in apparel, household fabrics, and military goods. The major market

TABLE 3

Properties of Repellent Finishes

Type	Resistance to		Relative cost ($) per lb[a] active solids	Concentrations in padding bath (%)	Repellency
	Laundering	Dry cleaning			
Nondurable wax	Moderate	Very poor	1.00	12	Water
Resin type	Good	Poor	1.00	10	Water
Silicone	Moderate	Good	3.00	6	Water
Pyridinium	Very good	Moderate	1.30	8	Water
Fluorochemical, extended	Very good	Good	22.00	3[a]	Water and oil
Fluorochemical-pyridinium (Quarpel)	Excellent	Good	22.00	15[a]	Water and oil

[a]Does not include extender or pyridinium salt, where applicable. These are added at a bath concentration of about 3 and 8%, respectively. The extender costs about $2 per pound of active solids.

for nondurable, wax-emulsion repellents is in recreation equipment. Note that the solids content of finishes is only part repellent. Other components are present which serve purposes complimentary to that of the repellent. Some chemicals which may be added to a fluorochemical repellent are given here as examples.

Extenders are durable water repellents which lower the overall cost of the finish. They are generally used at a level of 1.5 percent of the fabric weight. Extenders are durable water repellents of the resin, pyridinium, or wax types. Some tradenames are Nalan W (DuPont), Norane R (Sun Chemical), Permel B (Cyanamid), Phobotex FTC (Ciba) and Ranedare (Metro-Atlantic).

Hand Modifiers offset the effect of the fluorochemical on the hand of the fabric.

Thermosetting resins impart durable press or wash-and-wear properties to the fabric and improve the fastness of the repellent.

Silicone repellents and anionic auxiliaries adversely affect oil repellency and are not used with fluorochemicals.

A. Fluorochemical Finishes

1. Commercial Aspects

Fluorochemicals are the most popular of the repellent finishes since they repel water and oil- and water-borne stains, are very efficient and require little add-on, and are particularly durable to laundering and dry cleaning. However, they are expensive. These considerations explain that fluorocarbons account for almost 80 percent of the dollar volume and only 20 percent of the repellent consumed. It should be noted, however, that the solids content of the expensive fluorochemical finishes is often only partly fluorochemical. These finishes are supplied as two component systems -- fluorochemical and extender.

Table 4, which gives the consumption of finish by product, shows the remainder of the repellent product market is satisfied, for the most part, by waxes, silicones, and pyridinium salts. Table 5 gives the breakdown of consumption by use. Most of the repellents see use in areas such as automotive fabrics, recreational products (tents, sleeping bags), and apparel. However, style and taste can dramatically change the demand for finishes for apparel where the effect of the finish is not critical for function. Military consumption is analogously subject to changes in political situations.

Many companies supply repellent formulations. However, only a few sell the basic, active components of repellent finishes. DuPont and 3M, the major U.S. producers of fluorochemicals, sell directly to the textile industry. Pennwalt and Ciba-Geigy also produce fluorochemicals. About one quarter of the fluorocarbon repellents are applied from solvent. The rest of this class and all other repellent finishes are applied from aqueous emulsions. The solvents are themselves fluorocarbons and 1,1,1-trichloroethane or trichloroethylene. They are applied at a wet pickup of 60-100 percent to give an add-on of 0.5 percent solids. Table 6 describes the principle grades of fluorochemical repellents.

A typical formulation for polyester-cotton rainwear is:

Component	Concentration (%)
Fluorochemical	2.5
Extender	1.0
Catalyst	0.25
Thermosetting resin	5.0
Accelerator	1.0
Softener	1.0
Bath stabilizer	0.03
Organic acid	0.05

The mixture is padded on at room temperature, dried at 300°F and cured at 340°F, for one to three minutes, depending on the weight of the fabric.

TABLE 4

U. S. Consumption of Water and Oil Repellents in 1968

Repellent	Million lb.
Fluorochemicals	1. 0
Extenders	1. 0
Silicones	0. 5
Pyridinium salts	0. 25
Waxes	2. 0
Other	0. 25
Total	5. 0

TABLE 5

U. S. Consumption of Fluorochemicals in 1968 by Use

End use	Percent
Rainwear	25
Other apparel	30
Upholstery	25
Other home furnishings	5
Automobile fabrics	10
Other	5

Most fluorochemical finishes develop their maximum, initial oil and water repellency at 1 percent solids on the fabric [10]. When durability to laundering and dry cleaning is considered, concentrations of 2 percent are desirable. However, cost considerations will temper the finisher's desire to achieve this result.

One must be cautious not to generalize about treatment with fluorochemicals. It is best to tailor the treatment to the fabric. For example, the choice of an emulsifier should be guided by the need for the charge on the latex particle to be neutral or opposite from that of the net charge on the fabric. Anionic latexes are best for woolens at an equilibrium pH below the isoelectric point for wool. Cationic or nonionic latexes are best for cottons and for woolens at equilibrium pH's above their isoelectric point.

TABLE 6

Principle Grades of Fluorochemical Repellents[a]

Product designation	Solvent system	Method of Application	Uses and properties obtained[b]
Zepel B and DR	Cationic dispersion	Padding	Rainwear, outerwear, general apparel, cottons, drapery fabrics, upholstery.
Zepel D	Cationic dispersion	Padding	Rainwear, outerwear, military finishes; approved for Quarpel.
Zepel K	Cationic dispersion	Padding	Outerwear, durable press and easy care apparel. Needs only low temperature drying after washing to restore repellency.
Zepel RT	Cationic dispersion	Padding	Textured polyester and other synthetics.
Zepel 2829 and 2952	Cationic dispersion	Padding	Home furnishings, upholstery. Resistance to dry soiling and to liquids and to removal of repellency by abrasion.
Zepel velvet protector	Cationic dispersion	Padding	Cotton and rayon velvets.
Multi action Zepel	Cationic dispersion	Padding	100% polyester fabrics.
Zepel 2979	Solution in fluorocarbon and 1,1,1-trichloroethane	Padding, spraying	Home furnishings, woolens, upholstery, fabrics not suitable for aqueous processing.
TLF 3195 and 3196	Built, fluorocarbon/wax, cationic dispersion	Padding	Nonwovens.
TLF 2248, 3574 and 3580	Cationic dispersion	Padding	Nonwovens.

Product	Composition	Application	Uses
Nalan GN and W and TLF 3039	Polymer/wax dispersion	Padding	Durable water repellent. Adjuvant for fluorochemical finishing of nonwovens.
Scotchgard FC 208	Cationic dispersion	Padding	Cotton, rayon and synthetic blends for rainwear.
Scotchgard FC 210	Nonionic dispersion	Padding	Cotton and cotton blends; particularly for wash and wear, all white fabrics, rainwear. Approved for Quarpel.
Scotchgard FC 214A and B	Two part system, cationic dispersion	Padding, spraying	Upholstery. Good dry soil resistance.
Scotchgard FC 218	Cationic dispersion	Padding	Polyester/cotton and 100% synthetic, permanent press apparel and home use fabrics.
Scotchgard FC 232	Cationic dispersion	Padding	100% synthetic or synthetic blend, non resin treated fabrics for rainwear and outerwear.
Scotchgard FC 234	Cationic dispersion	Padding, spraying	Cellulosic pile upholstery fabrics such as velvets.
Scotchgard FC 321	Chlorinated or fluorinated solvents	Padding, spraying	Upholstery fabrics.
Scotchgard FC 324	Chlorinated or fluorinated solvents	Padding, spraying	Cellulosic pile upholstery fabrics such as woven and tufted velvets.
Scotchban FC 805	Cationic dispersion of Cr complex in water/alcohol	Padding	Nonwovens. Repellency to and hold out of liquids including oil, water and low surface tension solutions.
Scotchban FC 808 and 824	Cationic dispersion in water or water/alcohol	Padding	Nonwovens.

aZepel, TLF and Nalan are tradenames of the DuPont Company. Scotchgard and Scotchban are tradenames of the 3M Company.

bUnless otherwise noted, the products are used as oil and water repellents.

Curing conditions that are best to achieve optimum oil repellency may not be optimum for water repellency. Highest oil-repellency ratings (3M test) of 120, and a moderate water-repellency rating of 70, were obtained with a cotton print cloth treated with FC-208 and cured by air drying. Further curing at 300°F developed an excellent water-repellency rating of 100. However, the oil-repellency rating dropped to 104. Different results are obtained with other fabrics such as rayon, wool, or synthetics. Generally, both water and oil repellency improve with heat curing.

Although fluorochemicals impart excellent water repellency, the addition of some water repellents help stabilize the fluorochemical-emulsion polymers against shear-agitation breakdown in the pad box — and a synergistic enhancement of repellency properties is seen (especially that of oil repellency). It should be noted that most silicone water repellents or softeners reduce the oil repellency of fluorochemicals, even when present in trace amounts.

Fluorochemical-pyridinium repellents [45] are synergistic and most effective only in certain ratios of the two components (Table 7). Durability to laundering and dry cleaning increases with increasing concentration of fluorochemical resin. The Quarpel finish for military fabrics is such a treatment. A typical formulation, requiring a minimum of 1.0% fluorochemical, is:

fluorochemical (28% solids)	7.2%
pyridinium salt	8.0%
anhydrous sodium acetate	0.8%

TABLE 7

Performance of Fluorochemical and Pyridinium Repellents[a]

Repellent, % solids on fabric	Initial		Five dry cleanings		Five launderings	
	Oil	Spray	Oil	Spray	Oil	Spray
Fluorochemical, 0.4	100	100	70	70	50	90
Pyridinium, 2.5	0	0	0	0	0	100
Fluorochemical, 0.4 + pyridinium, 2.5	70	100	50	0	50	100
Quarpel	100	100	100	90	90	100

[a]3M ratings.

The synergism in effecting durable water and oil repellency, that is exhibited by the Quarpel finish, is not only due to the presence of the pyridinium water repellent. It is due in great degree to the high concentration (about 2 percent) of both fluorochemical and pyridinium salt, and also to the fact that the pyridinium salt is fiber reactive.

2. Structure of Fluorochemical Repellents

The fluoroacrylic esters (I) are examples of the fluorochemical component of good, commercial repellents. R_f is a perfluorinated alkyl group in the formula.

$$\left[\begin{array}{l} \text{CHCH}_2\!\!- \\ \quad | \\ \text{COO(CH}_2)_{1,2}\text{R}_f \end{array} \right]_n$$

(I)

The perfluoroacrylates are generally made from straight-chain per-fluorinated alcohols [46, 23]. When a fluoroketone, like hexafluoro-acetone is the starting material, the resulting fluorocarbon chain is capped with a perfluoroisopropoxy group [47].

Textile fibers and fabric surfaces do not present ideal surfaces as used by Zisman in his research. Perfluoro compounds, which show critical surface free energy values, γ_c, of 6 dynes/cm on smooth, clean, nonporous surfaces, are usually wet by hydrocarbons having a surface tension of 22 dynes/cm or less. The rough surface prevents optimum packing of the terminal CF_3 groups. A shift in wettability to that of the exposed, underlying atoms is effected.

Weak, intermolecular forces of fluorocarbons impart low free energies to their surfaces. Monolayers composed of closely packed fluorocarbon derivatives, whose fluorocarbon moieties are oriented in the same outward direction, also exhibit a low surface free energy. In this way, highly fluorinated organic solids have surfaces which are the least wettable and adherable known and thereby are both oil and water repellent [48-51].

Replacement of hydrogen in a polyethylene surface by fluorine, as in polytetrafluoroethylene, gives a value of 18 dynes/cm. The mini-mum value for γ_c was found to be 6 dynes/cm for perfluorolauric acid, $C_{11}F_{23}COOH$. Replacement of a terminal fluorine by hydrogen in a long-chain perfluoroacid increased γ_c to 15 dynes/cm. Long chains are needed to compensate for the large dipole of the CF_3 group (1.9 D) and to achieve closest packing. This is accomplished with the C_8F_{17} group. Most commercial fluorochemical surface modifiers have a C_7F_{15} or C_8F_{17} group and are in fact mixtures of these and other homologues. Larger perfluoroalkyl groups adversely affect solubility and other properties.

Branching of a perfluoroalkyl chain disrupts close packing. Thus, $(CF_3)_2CF(CF_2)_{11}COOH$ has a γ_c of 13.3 dynes/cm -- more than twice that of the straight-chain acid. Branching and shorter length in $(CF_3)_2CFCF_2COOH$ result in a γ_c of 15.2 cynes/cm. The straight-chain C_5 perfluoroacid has a γ_c of < 10 dynes/cm. An ether link buried in a perfluoroalkyl group does not affect γ_c [52, 53]. In light of the evidence that branching is less desirable for repellency, it is surprising to note that the perfluorisopropoxy group, $(CF_3)_2CFO-$, was found to provide the oleophobicity of 6-7 fluorinated carbons in a straight chain [54, 55].

The part of the molecule to which the perfluoro group is attached also affects the packing of that group and the oil or water solubility. Thus, at 1% concentration in water, $C_7F_{15}COOH$ and $C_8F_{17}SO_3H$ reduce the surface tension of water to 15 and 30 dynes/cm respectively.

A fluorinated chain at least four carbons long, capped by a CF_3 group, is necessary to produce good oil repellency. However, it is not sufficient. The type and nature of the bond between the perfluoro group and the substrate influences not only the repellency but the durability of the treatment. A bonding group which produces a tacky film will lower the resistance of the finish to dry soiling. Tacky films are also obtained when low-molecular-weight fluorocarbon homologues are used in the synthesis of the repellent and when waxy extenders are used. A hydrophilic bonding group can offset the repellent effect of the oriented perfluoro group. This property forms the basis of a soil-releasing finish made by the reaction of ethylperfluorooctanoate and ethylenimine to form a perfluoroacylaziridine which in turn polymerizes spontaneously at room temperature or by heating [56]. The polymer can react with functional groups on the fiber such as hydroxyl and carboxyl [57]. High oil repellency, but little water repellency, results. The oil repellency is stable to laundering and to dry cleaning. The poor water repellency is due to the relatively hydrophilic amido groups. Similar results were obtained with the reaction product of 1,1-dihydroperfluorooctylamine and tetrakis (hydroxymethyl) phosphonium chloride which is cured onto the fabric with ammonia [58, 59].

Water repellency is not as sensitive to chain length as is oil repellency. A rather constant water repellency value of 70 is obtained

TABLE 8

Repellency on Cotton Cloth for a Series of Poly (fluorocarbon acrylates), R_fCH_2O-acrylate [10]

R_f	Oil repellency[a]	Water repellency[b]
CF_3	0	50
C_2F_5	60	70
C_3F_7	90	70
C_5F_{11}	100	70
C_7F_{15}	120	70
C_9F_{19}	130	80

[a]3M oil repellency test.
[b]AATCC spray test, Method 22-1952.

when the chain length exceeds C_2F_5, whereas oil repellency increases from 0 to 130 as the chain length increases from CF_3 to C_9F_{19}. Table 8 shows this for a series of poly-1,1-dihydroperfluoroacrylates on cotton [10]. This reflects the more stringent requirements of oils which have much lower surface tensions than water and therefore wet surfaces having relatively low values of surface free energy.

3. A Survey of Fluorochemical Finishes

The perfluorocarboxylic acids are the main starting materials for fluorocarbon repellents. These are made by direct fluorination of the hydrocarbon analogs [60]. Tetrafluoroethylene and its telomer iodides are also important starting materials. Chromium coordination complexes of these acids and of polymers of fluoroalkyl esters of acrylic acid show excellent oil and water repellency, but are not durable to alkaline washes or to dry cleaning [11, 61]. Partial esters of cellulose, substituted with perfluorocarboxyl groups are durable to neutral detergents and to dry cleaning. However, they hydrolyze in the presence of acid or base [62].

Copolymers of the perfluoroalkylacrylates and methacrylates, and their copolymers with other monomers, have been used to achieve

different effects. Examples are the copolymers of acryloyl and meth-
acryloyl chloride with 1,1-dihydrofluorooctyl acrylate to give water
and oil repellents [63-65]. Polyfluoralkylalkenylalkoxyacrylates,
$R_fCH=CHCH_2OCH_2CH_2OOCCH=CH_2$, provide both oil repellency and
oil release after staining [66]. Perfluoralkylethylmethacrylates,
$R_fCH_2CH_2OOCC(CH_3)=CH_2$, are claimed to give repellency to dry soil
as well as to oil and water [67]. Copolymerization of the latter with
N-methylolacrylamide and 2-hydroxyethylacrylate yields a more durable product with the advantage of easy care [68]. This was also previously obtained when easy-care ingredients were added to a copolymer
of a perfluoroalkylacrylate and trifluorovinyl ether [69]. When the
fluorocarbon esters are copolymerized with glycidyl acrylates amd
methacrylates, one obtains trichloroethylene-soluble, dry-soil-re-
sistant finishes [70]. When the fluorocarbon acryloyl chlorides are
reacted with ethylene or propylene oxides, in the presence of an alk-
oxide, acrylates which contain one or more alkyleneoxy groups are obtained. The latter groups are relatively hydrophilic and provide soil
release [71].

The order of addition of protective treatments is important. One
patent teaches the preliminary curing of a fluorochemical compound
which resists the migration of a solvent-borne, waterproofing resin
which may close the pores of the fabric [72]. A later patent prefers
first treating the fabric with the resin, an ethylene-vinyl acetate copolymer, in an aqueous dispersion. Subsequent addition of a water-
immiscible solvent, such as trichloroethylene, swells the copolymer
of the dispersion so that it promotes preferential subsequent coating
of the fabric on the surface. Then the fluorocarbon is padded on and
cured [73].

A variety of functional groups, and hetero-atom-containing linkages, have been used for reasons such as reactivity with the textile
surface, molecular compatibility, similarity with that of the textile,
and addition of some degree of hydrophilicity for soil release. Some
groups appearing in fluorocarbon repellents have been amino, sulfido,
amido, thioamido, dithioamido, ester, thioester, dithioester [74],
sulfonamido [75], carbamyl, isocyanato [76], trialkoxysilyl [77] and
oxazolinyl [78]. Reaction of a fluorocarbon amine with a tetrahydroxy-
methyl phosphonium salt gave a repellent particularly reactive to cel-
lulosics and which could be applied from aqueous emulsions [79]. Introduction of alkyleneimino [80] or alkyleneoxy [81] groups results in
increased hydrophilicity. This was accomplished by copolymerization
of oleophobic fluorocarbon silanes with hydrophilic alkyleneoxy silanes
[82]. Hydrophilicity was also obtained with perfluoroesters of polymeric phosphonitrilic acid [83] and with the reaction products of perfluoroalkyl iodides and telomer ester adducts of diethylvinyl phosphon-
ate [84].

The past few years have seen a flurry of activity to synthesize derivatives which contain the perfluoroisopropoxy group. It would appear that the main purpose of such synthesis is to avoid infringement on patents which are based on straight-chain fluorocarbons. However, in a description of the synthesis of acrylates, the hexafluoroisopropyl radical was said to have oleophobicity equivalent to six or seven fluorinated carbon atoms in a straight chain [55]. An early patent describes the synthesis of 1,4-bis-(heptafluoroisopropoxy)-2-butene whose epoxide is converted into polymers [85, 86].

The starting materials for perfluoroisopropoxy derivatives are fluoroketones (hexafluoroacetone being the most important), tetrafluoroethylene, an ionizable fluoride and iodine [86-89].

$$(R_f)_2 CO + KF \longrightarrow (R_f)_2 CFOK \tag{3}$$

$$(R_f)_2 CFOK + (n+1)C_2F_4 + I_2 \longrightarrow (R_f)_2 CFO(CF_2CF_2)_n CF_2CF_2I \tag{4}$$

$$(R_f)_2 CFO(CF_2CF_2)_n CF_2CF_2I \xrightarrow[\text{2. MeOH}]{\text{1. SO}_3} (R_f)_2 CFO(CF_2CF_2)_n CF_2COOMe \tag{5}$$

where R_f denotes a perfluorocarbon radical such as CF_3. The esters can be converted to derivatives such as amides or reduced to alcohols which in turn can be esterified. The fluoroalkyl iodide has been added across the double bond of unsaturated alcohols, nitriles, acids, and esters; and the iodine has been subsequently reduced to hydrogen. The carboxylic acids readily form chromium complexes which can be applied to paper and leather [90, 91]. The fluoroalkyl alcohols can be converted to carbamates, dicarbamates [92], and isocyanates [93]. The monocarbamates show more repellency and ability to withstand laundering than do the dicarbamates [92]. A method used to produce a compound loaded with perfluoroalkyl groups is one where a fluorocarbon ester is allowed to react with a polyfunctional compound. For example, the perfluoroisopropoxy fluoroalkyl ester reacts easily with the two primary amino groups of diethylenetriamine. The secondary amino group remains to enable reaction with other polyfunctional compounds [94].

In addition to polyamides, also polyfunctional ureas, esters, isocyanurates and isocyanates have been produced. Some of these compounds, called amphipathic additives, have the fortunate property of being partially soluble or dispersible in the polymer of which the textile is made [95]. This affinity is due to the nonfluorocarbon part of the additive. The fluorocarbon part provides the repellency. If the compound is sufficiently thermally stable (200-350°C) to survive melt-processing and extrusion temperatures, it can be blended in with the resin mixture before extrusion. The extruded filament has a repellent surface which is substantive to scouring, laundering, and dry cleaning. If, in addition, the nonfluorocarbon portion is of the correct structure, the major part of the additive migrates toward the surface of the filament, thereby making good use of most of the additive. Drawing the filament to a finer denier causes a spreading of the additive over a larger surface with a concomitant reduction in repellency. However, annealing the drawn filament or the textile promotes sufficient migration of additive from the filament core to the surface to reestablish repellency. In fact, the conditions of scouring of the textile are often sufficient to achieve this [96]. Such treatments affect neither the hand nor the dyeability of the textile.

B. Silicone Finishes

Stable, methyl-substituted polysiloxanes (II) (R = CH_3, H) bond strongly to a fabric. Condensation can occur with hydroxyl groups, which are part of the fiber molecules or with the moisture film on the fiber surface. Silicones for textiles, however, are for the most part mixtures of both polymethylsiloxanes and polymethylhydrogensiloxanes where the hydrogens condense with the functional groups on the textile surface [43, 97]. Increased resistance to laundering and to dry cleaning is obtained by substituting polydivinylsiloxane for the polydimethylsiloxane [98] and by using an even more reactive siloxane containing silanol groups [99]. Both methods provide crosslinking sites - the first through the vinyl groups and the second through the hydroxyl groups.

$$
\begin{array}{c}
CH_3 \\
| \\
- Si\text{-}O - \\
| \\
R
\end{array}
$$

(II)

Silicones are padded on from aqueous emulsion baths containing 0.3-0.5 percent solids. Metal soap catalysts are used to fix the finish. The amount of catalyst varies between 4-8 percent of the dry weight of the silicone. Drying and curing are done at 230-350°F. The fabric is washed, when easy-care resins are used, to remove unfixed resin, acid and other residues. Most commercial products require high-temperature curing whether a catalyst is used or not. Room-temperature curing can be obtained with silanol-terminated products together with a crosslinking agent such as ethyl silicate in an organic solvent. The catalyst is a tin soap or an organotitanium compound [100]. Curing can be activated by atmospheric moisture if hydrolyzable polyfunctional silanes or siloxanes are used as crosslinking agents.

The repellency of silicones is thought to reside in the methyl groups oriented along the fiber surface rather than in long chains, normal to the surface, as in the fluorochemical or hydrocarbon repellents. In fact, the maximum water-repellent effect is obtained when the hydrocarbon chain is about sixteen units long. Longer chains tend to coil, and shorter chains do not insulate the terminal polar groups well.

The properties obtained with silicones are durable water repellency; resistance to water-borne stains; improved tear strength; abrasion resistance; sewability; and a soft, slick hand. However, the silky feel is not always desirable. Silicones find their largest applications in rainwear, outerwear, sportswear, pile fabrics, and home furnishings.

Consumption of silicones is declining in favor of fluorochemicals which repel both stains and water. Air-borne silicone droplets also cause problems in mills due to the slipperiness they impart to surfaces. Most silicones (70%) for the textile industry are manufactured by Dow-Corning and sold neat or in the form of emulsions. General Electric also produces silicones.

Somewhat related to the silicones is a treatment consisting of silica particles and a multivalent metal which renders the fabric soil resistant [101]. Also, a novel treatment called Silanox 101 was recently disclosed [102]. The product is a pyrogenic silica which has been surface-treated with a water-repellent chemical whose composition is disclosed only in that it covers the silica with methyl groups. The material is so hydrophobic that aqueous dispersions could only be effected by first encapsulating it in air and then the air-encapsulated material in water. This was accomplished with Carbopol 941, an air-water emulsifying agent. Polytetrafluoroethylene and paraffin-wax surfaces which had contact angles with water of 98° and 68°, respectively, showed contact angles of 127° and 143° after treatment. Water shedding is extremely efficient in that a Silanox 101-coated Kraft paper rids itself of a water droplet after being tilted only 1° from the horizontal.

The nature of the surface is such that it should be similar to that of paraffin. However, since the water contact angle is greater than 90°, theory predicts an even greater contact angle on a rough surface than is obtained with this particulate surface. An untoward effect of the particulate nature of the treatment is that it is discontinuous and not film-like. Thus, it leaks under pressure, failing the Mason jar test under a static head of only 0.5 in. It would seem that its resistance to laundering would also be severely limited.

C. Waxes and Metal Salts

Most wax treatments are emulsions of paraffin wax containing a metal salt such as a basic aluminum acetate or formate. Zirconium salts, which now account for half the market for these treatments, impart greater durability to laundering. However, even these do not stand up to laundering or dry cleaning as well as do the more durable finishes; neither do they resist oily stains. The initial attraction of these waxes were their ease of use, inexpensiveness, and broad utility. Due to their transient adherence, they are mainly used on cotton awnings, tents, and boat covers. Few are used on synthetic fibers, except as an agent which masks the dirt by reflecting light from its surfaces. Good repellency is obtained at an add-on level of the wax emulsions of 2-6 percent of the fabric weight. Padding is done at 130-160°F. All emulsions are maintained in neutral or slightly acid state, since they are unstable in alkali.

A popular modification, twenty years ago, was a single bath treatment with an emulsion of paraffin wax. Emulsions without metal salts are still used on hydrophobic, filament yarns such as nylon or polyester where precipitation within a fiber is not needed. At the beginning of the development of the metal-salt systems, aluminum acetate was precipitated when aluminum sulfate was allowed to react with lead acetate. If the aluminum acetate was made alkaline with sodium hydroxide, hydrated alumina was precipitated. Later, a soap was substituted for sodium hydroxide. The precipitate was then an aluminum soap -- usually aluminum stearate. Better penetration of the fiber was obtained with a water-insoluble, basic aluminum salt having an ultimate particle size of less than 0.5 microns [103].

Zirconium salts replaced aluminum salts because their lower solubility imparted greater substantiveness to laundering and to dry cleaning [104, 105]. Thus, zirconium stearate, on reaction with caustic, remains insoluble as a hydroxide or a carbonate; whereas the amphoteric aluminum salts dissolve in caustic. A sodium acetate buffer is necessary with zirconium oxychloride to neutralize the damaging hydrogen chloride which is evolved on hydrolysis.

Soaps of thorium, uranium, and the rare earths [104], as well as tetraalkyl orthotitanates [106], have been claimed. A mixture of two hydrous, stable metal oxides is also suggested [107]. Organic solutions permit the use of alkali titanates to prevent dry soiling. A reaction product of salts of titanium, aluminum, or zirconium and an alkyl-substituted succinic anhydride or its monoester [108] is suggested as a water repellent which has a low tendency to form gels. A coordination complex, whose formula corresponds to hydroxostearato-dichromium(III) chloride, is sold by DuPont as Quilon. The fabric is treated with a solution of this compound at 120°C, where it rearranges during hydrolysis to give a water repellent surface of organic rings oriented by chelated chromium [109, 110]. Other hydrocarbon chromium compounds are also important water repellents.

D. Pyridinium Salts and Other Nitrogen Compounds

1. Pyridinium Salts

These water repellents are derivatives of the water soluble, fiber penetrating, stearamidomethyl pyridinium chloride (III). They are made by heating the amide and formaldehyde together in pyridine [111]. Padding, curing, and thereby insolubilization, is performed at 300-350°F. During curing, decomposition occurs to give formaldehyde, pyridine, and hydrochloric acid and the fiber-reactive, water-insoluble stearamidomethyl radical. This radical can form a hemiacetal with the hydroxyl groups of cellulose or it can precipitate as N, N'-methylene-bis-stearamide (IV) [112]. Although formulations are buffered with sodium acetate to neutralize the hydrochloric acid, the treated fabric is further washed with hot alkali to remove residual acid and other decomposition products. It is then rinsed thoroughly and dried at 275-350°F.

$$CH_2(CH_2)_{16} CONHCH_2Py^+Cl^- \qquad\qquad [CH_3(CH_2)_{16} CONH]_2CH_2$$

$$(III) \qquad\qquad\qquad\qquad\qquad (IV)$$

To effect good initial water repellency, an add-on based on 2-6 percent of fabric weight is required. Larger concentrations produce stiff finishes which show chalk marks, especially on smooth-filament fibers. Most pyridinium repellents are applied to rainwear, and in combination with fluorocarbons, since by themselves they have no stain resistance and are not durable to dry cleaning. When used alone, the pyridinium repellent that remains after laundering is revealed by ironing. It is theorized that ironing melts the waxlike material and redistributes it over the surface. DuPont, the major supplier (about 75% of the market) of pyridinium repellents, markets them under the name of Zelan.

2. Other Nitrogen-Containing Repellents

Nitrogen-containing compounds can be fixed by making them less soluble through reaction with the basic nitrogen. Thus, glue was allowed to react with aluminum compounds or with dichromates [113]. Some formulations are similar to permanent press resins. Thus, glue has been condensed with formaldehyde and with melamine. Urea-formaldehyde resins have also been used [114]. N-methylolacrylamide has been used, with and without an aldehyde condensate, to crosslink with cellulose [115].

The repellent 1-n-octadecyl-3-ethylene urea is claimed to be superior to stearamide for rayon. It is made by reacting n-octadecyl isocyanate with ethyleneimine in emulsion [116]. A polyisocyanate has also been reacted with a polyol and then with a polymeric acid to give a water repellent [117]. Most hydrophobic finishes that coat fabric will show repellency. Thus, antistatic and soil resistance properties are obtained with the polyester formed by reacting a polymeric acid, containing carboxyl, sulfonic acid, or dihydrogen phosphate groups, with a polyol or with a compound containing epoxide groups [117]. Similarly, water repellency is claimed for a mixture of mineral oil, base cordage oil, oleic acid, and a cationic wetting agent such as trimethyl-amidoethyl ammonium sulfate [118].

REFERENCES

1. F. W. Minor and A. M. Schwartz, Text. Res. J., 29, 931 (1959).
2. F. W. Minor and A. M. Schwartz, Text. Res. J., 29, 940 (1959).
3. R. E. Johnson, Jr. and R. H. Dettre, Advances in Chemistry Series, No. 43, pp. 112-144, American Chemical Society, Washington, D.C., 1964.
4. R. E. Johnson, Jr. and R. H. Dettre, J. Phys. Chem., 68, 1744 (1964).
5. R. H. Dettre and R. E. Johnson, Jr., J. Phys. Chem., 69, 1507 (1965).
6. C. G. L. Furmidge, J. Colloid Sci., 17, 309 (1962).
7. J. Berch, H. Peper and G. L. Drake, Jr., Text. Res. J., 35, 252 (1965).
8. S. Baxter and A. B. D. Cassie, J. Text. Inst., 36, T67 (1945).
9. R. J. Berni, R. R. Benerito, and F. J. Philips, Text. Res. J., 30, 576 (1960).
10. E. J. Grajeck and W. H. Petersen, Text. Res. J., 32, 320 (1962).
11. F. J. Philips, L. Segal, and L. Loeb, Text. Res. J., 27, 369 (1957).
12. R. E. Read and G. C. Culling, Amer. Dyestuff Reporter, 56, 881 (1967).
13. K. Shinoda, "Solvent Properties of Surfactant Solutions," Dekker, New York, 1967, p. 157.
14. B. Miller, A. B. Coe, and P. N. Ramacandran, Text. Res. J., 37, 919 (1967).
15. R. Tsuzuki and N. Yabuuchi, Amer. Dyestuff Reporter, 57, 472 (1968).
16. C. B. Brown, S. H. Thompson, and G. Stewart, Text. Res. J., 38, 735 (1968).
17. J. Compton and W. J. Hart, Text. Res. J., 23, 158 (1953).
18. J. Compton and W. J. Hart, Text. Res. J., 23, 418 (1953).
19. J. Compton and W. J. Hart, Text. Res. J., 24, 263 (1954).

20. W. Kling and H. Mahl, Melliand Textilber., 35, 640 (1954).
21. B. R. Porter, et. al., Text. Res. J., 27, 833 (1957).
22. V. W. Tripp, et. al., Text. Res. J., 28, 447 (1958).
23. V. W. Tripp, R. L. Clayton, and B. R. Porter, Text. Res. J., 27, 340 (1957).
24. T. Fort, Jr., H. R. Billica, and C. K. Sloan, Text. Res. J., 36, 7 (1966).
25. J. M. Collins, O. C. Bacon, and J. E. Smith, Am. Dyestuff Reporter, 51, 20 (1962).
26. G. A. Slowinske, Am. Dyestuff Reporter, 30, 6 (1941).
27. G. A. Slowinske and A. G. Pope, Am. Dyestuff Reporter, 36, 190 (1947).
28. H. G. Goldstein, Text. Res. J., 31, 377 (1961).
29. AATCC 118-1966T Test Method, Oil Repellency, Hydrocarbon Resistance Test, "AATCC Technical Manual," 1967, p. B-139.
30. E. R. Kaswell, "Water Repellency and Water Resistance in Textile Fibers, Yarns and Fabrics," Reinhold, New York, 1953, pp. 236-255.
31. G. E. Martin, H. S. Sell, and B. W. Habeck, Rubber Age (N.Y.), 66, 409 (1950).
32. G. E. Martin, H. S. Sell, and B. W. Habeck, Text. Res. J., 20, 123 (1950).
33. L. Fourt and N. R. S. Hollies, "Clothing Comfort and Function," Dekker, New York, 1970, Ch. 6.
34. Textile Institute Tentative Specifications Nos. 7, 8, J. Text. Inst., 38, 178 (1945).
35. F. T. Peirce and W. C. Gardiner, U.S. Pat. 2,350,696 (1944).
36. A. M. Sookne, The Problem of Water Repellent Fabrics, in "Symposium on the Functional Properties of Clothing Fabrics," Textile Research Institute, New York, 1943, pp. 12-15.
37. J. D. Reid and C. F. Goldthwait, U.S. Dept. Agr. Yearbook, 1950-51, pp. 411-418.
38. C. F. Goldthwait and H. O. Smith, Textile World, 95, 105 (1945).
39. M. Mayer, Jr., G. J. Kyame, and J. J. Brown, Textile World, 102, 114 (1952).
40. Mod. Text., 44 (28), 85 (1963).
41. J. R. Caldwell and C. C. Dannelley, Amer. Dyestuff Reporter, 56, 77 (1967).
42. J. M. May, Amer. Dyestuff Reporter, 58, 15 (1969).
43. R. L. Wayland, Jr., et. al., Amer. Duestyff Reporter, 52, 17 (1963).
44. A. A. Schuyten, J. W. Weaver, and J. D. Reid, Amer. Dyestuff Reporter, 38, 364 (1949).
45. C. G. De Marco, A. J. McQuade, and S. J. Kennedy, Mod. Text., 2, 50 (1960).
46. F. A. Bovey, et. al., J. Polym. Sci., 15, 520 (1955).
47. J. P. Moreau, S. E. Ellzey, Jr., and G. L. Drake, Jr., Text. Res. J., 56, 117 (1967).

48. W. C. Bigelow, D. L. Pickett, and W. A. Zisman, J. Colloid Sci., 1, 513 (1946).

49. A. H. Ellison, H. W. Fox, and W. A. Zisman, J. Phys. Chem., 57, 622 (1952).

50. H. W. Fox and W. A. Zisman, J. Colloid Sci., 7, 109, 428 (1952).

51. F. Shulman and W. A. Zisman, J. Colloid Sci., 7, 465 (1952).

52. T. J. Brice, W. H. Pearlson, and H. M. Scholberg, U. S. Pat. 2,713,593 (1955).

53. F. A. Bovey and J. F. Abere, U.S. Patent 2,826,564 (1958).

54. A. G. Pittman and W. L. Wasley, U.S. Pat. 3,574,713 (1971).

55. A. G. Pittman and W. L. Wasley, U.S. Pat. 3,701,792 (1972).

56. A. G. Pittman, Text. Res. J., 33, 953 (1963).

57. J. P. Moreau and G. L. Drake, Jr., Amer. Dyestuff Reporter, 58, 21 (1961).

58. S. E. Ellzey, Jr., et. al., Text. Res. J., 39, 809 (1969).

59. S. E. Ellzey, Jr., et. al., U.S. Pat. 3,701,626 (1972).

60. E. A. Kauck and A. R. Dresslin, Ind. Eng. Chem., 43, 2332 (1950).

61. L. Segal, F. J. Philips, L. Loeb, and R. L. Clayton, Jr., Text. Res. J., 28, 233 (1958).

62. R. R. Benerito, R. J. Berni, and T. E. Fogley, Text. Res. J., 30, 393 (1960).

63. A. G. Pittman and W. L. Wasley, U.S. Pat. 3,698,856 (1972).

64. A. H. Ahlbrecht, T. S. Reid, and D. R. Husted, U.S. Pat. 2,642,416 (1953).

65. A. H. Ahlbrecht and S. S. Smith, U.S. Pat. 3,102,103 (1963).

66. E. G. Stump, Jr., P. D. Schuman, and P. Tarrant, U.S. Pat. 3,625,929 (1971).

67. S. Raynolds, U.S. Pat. 3,645,990 (1972).

68. Brit. Pat. 1,261,152 (1972).

69. T. K. Tandy, Jr., U.S. Pat. 3,546,187 (1970).

70. E. J. Greenwood, U.S. Pat. 3,637,614 (1972).

71. A. G. Pittman, U.S. Pat. 3,654,244 (1972).

72. V. C. Smith, E. H. Hinton, and D. A. Davis, U.S. Pat. 3,326,713 (1967).

73. O. R. Crabtree and M. A. Thomas, U.S. Pat. 3,649,344 (1972).

74. C. S. Rondestvedt, Jr., U.S. Pat. 3,655,732 (1972).

75. P. C. V. Bouvet, C. M. H. E. Brouard, and J. P. C. Lalu, Brit. Pat. 1,283,266 (1972).

76. E. Schuierer, W. Renz, and H. Sommer, U.S. Pat. 3,679,634 (1972).

77. To Nalco Chemical Co., Brit. Pat. 1,267,224 (1972).

78. D. G. Gagliardi, U.S. Pat. 3,677,812 (1972).

79. S. E. Ellzey, Jr., W. J. Connick, W. A. Reeves, and G. L. Drake, Jr., U.S. Pat. 3,655,413 (1972).

80. W. J. Connick, Jr. and S. E. Ellzey, Jr., U.S. Pat. 3,655,435 (1972).

81. D. K. Ray-Chaudhuri and C. P. Iovine, U.S. Pat. 3,660,360 (1972).

82. A. G. Pittman and W. L. Wasley, U.S. Pats. 3,639,156 (1972), 3,702,859 (1972), 3,716,518 (1973), 3,716,519 (1973).

83. R. F. Stackel and G. P. Depaolo, U.S. Pat. 3,718,501 (1973).

84. L. H. Chance and J. Moreau, U.S. Pat. 3,719,448 (1973).

85. A. G. Pittman and W. L. Wasley, U.S. Pat. 3,653,956 (1972).

86. M. Litt, et. al., U.S. Pat. 3,453,333 (1969).

87. F. W. Evans and M. H. Litt, U.S. Pat. 3,470,256 (1969).

88. L. G. Anello, R. F. Sweeney, and M. H. Litt, U.S. Pat. 3,514,487 (1970).

89. L. G. Anello, R. F. Sweeney, and M. Litt, U.S. Pat. 3,697,564 (1972).

90. L. G. Anello and R. F. Sweeney, U.S. Pat. 3,644,454 (1972).

91. L. G. Anello, et. al., U.S. Pat. 3,651,105 (1972).

92. L. G. Anello and R. F. Sweeney, U.S. Pat. 3,657,320 (1972).

93. J. J. Murray, U.S. Pat. 3,657,306 (1972).

94. R. F. Sweeney and A. K. Price, U.S. Pat. 3,576,019 (1971).

95. B. C. Oxenrider and C. Woolf, Brit. Pat. 1,296,426 (1972).

96. B. C. Oxenrider and C. Woolf, U.S. Pat. 3,646,153 (1972).

97. F. L. Dennett, U.S. Pat. 2,588,365 (1952).

98. C. J. Conner, et. al., Text. Res. J., 30, 171 (1960).

99. J. W. Gilkey, Text. Res. J., 33, 129 (1963).

100. G. W. Madaras, J. Soc. Dyers Colour., 74, 835 (1958).

101. J. Schappel, U. S. Pat. 2,999,774 (1961).

102. R. F. Heitmuller, et. al., Amer. Dyestuff Reporter, 73, 75 (1973).

103. E. S. Pierce, S. S. Slowata, and S. J. O'Brien, U.S. Pat. 3,089,778 (1963).

104. C. B. White, U.S. Pat. 1,536,254 (1925).

105. J. J. Hirshfeld and B. J. Reuben, U. S. Pat. 3,671,292 (1972).

106. R. J. Speer and D. R. Carmody, Ind. Eng. Chem., 42, 251 (1950).

107. P. A. Florio and L. W. Rainard, U.S. Pat. 2,734,835 (1956).

108. F. Josten and W. Lucker, U.S. Pat. 3,684,566 (1972).

109. R. K. Iler, Ind. Eng. Chem., 46, 766 (1954).

110. R. K. Iler, U.S. Pat. 2,273,040 (1942).

111. A. W. Baldwin and E. E. Walker, Brit. Pat. 475,170 (1937).

112. H. A. Schuyten, et. al., Text. Res. J., 22, 424 (1952).

113. J. T. Marsh, "An Introduction to Textile Finishing," Chapman and Hall, London, 1966, pp. 458-494.

114. A. Morel, Ind. Text., 67, 450 (1950).

115. J. L. Gardon, U. S. Pats. 3,125,405 (1964) and 3,246,946 (1966).

116. M. A. Dahlen and R. A. Pingree, German Chemical Developments in Improving Water Resistance of Textiles, U. S. Dept. Commerce, Tech. Serv. P. B. Report, PB-1576, p. 91 (1945).

117. J. R. Caldwell and R. Gilkey, U.S. Pat. 3,236,685 (1966).

118. R. R. Matthews, U. S. Pat. 2, 876, 141 (1959).
119. Disposables Association Recommended Test (tentative) 80. 9-70
      (1970).

Chapter 14

SURFACE PROPERTIES OF GLASS FIBERS

Malcolm E. Schrader

Materials Department
David W. Taylor Naval Ship Research and Development Center
Annapolis, Maryland

I. INTRODUCTION

Glass, in the most general sense, consists of solid, noncrystalline materials. Since these materials lack a well defined melting point, they are sometimes considered liquids which are supercooled at room temperature [1]. The term, however, is most commonly used to refer specifically to the silicate class of glasses. These are inorganic, room-temperature glasses consisting of a three dimensional network of silicon and oxygen of empirical formula SiO_2, with various inorganic

additives which help determine their physical properties. For the case of the most common glasses, called "soft glass," or "soda-lime glass," the additives are alkali metal silicates. They disrupt the continuity of the silica network sufficiently to lower the softening point to a convenient region of approximately 400°C, at which the glass can be worked into a variety of desired shapes. Borosilicate glasses, known widely by the brand name "Pyrex," can consist, for example, of approximately 80% SiO_2, 13% B_2O_3, 4% Na_2O, 2% Al_2O_3, and 1% Li_2O and

K_2O. Their chief characteristic is high resistance to thermal shock,

which makes them uniquely suitable for a variety of applications, including laboratory glassware and kitchen cooking ware. The most common glass composition for drawing fibers for use in textiles or structural materials is a borosilicate known as "E glass" [2]. A typical E-glass composition would be, for example, 54% SiO_2, 17%

CaO, 15% Al_2O_3, 8% B_2O_3, and 5% MgO. A recent innovation has

involved drawing fibers from a melt of glass with the composition 65% SiO_2, 25% Al_2O_3, and 10% MgO, which yields a product with

especially high tensile strength known as "S glass" [3]. These find use in glass-fiber-reinforced plastics for high performance materials. The actual surface concentrations of the components of fibers will vary from the bulk composition according to the ability of each to lower the surface free energy by diffusing to the surface while the fiber is still in the molten state (surface segregation). Regardless of the actual composition of this fiber surface, however, it has been found that realistic approximations of the surface properties of borosilicates relevant to most uses are obtained by assuming that they are determined by the siliceous portion of the surface.

A theoretical, or ideal, glassy siliceous surface is that which would exist upon terminating the continuous, three-dimensional network of fused silica. This would result in surface exposure of an unknown proportion of silicon atoms with dangling bonds, oxygen atoms with dangling bonds, and siloxane groups $[(-Si-O-)_x]$.
The closest physical entity to this ideal construction would be a glass surface formed upon fracture in a vacuum. This could be

regarded as a surface resulting from an energetically-preferred cleavage of bulk-fused silica, followed by a possible surface rearrangement involving recombination of some of the dangling bonds. In practice, of course, the rearrangement would probably occur simultaneously with the fracture. This final vacuum-exposed surface would of necessity have many remaining dangling bonds and strained bonds or structures, in those areas affected by the rearrangements. These characteristics are components of the surface free energy which were available for reaction with suitable atmospheric ambients.

By far the most important atmospheric ambient, relevant to the surface of glass, is water vapor. Any site on the surface of fused silica can, in principle, react chemically with a water molecule. An oxysilicon radical could extract hydrogen from water while a silicon radical extracted a hydroxyl group (where Si_s is a surface silicon atom).

$$
\begin{array}{c}
\overset{\displaystyle \cdot O}{\diagup} \\
Si_s\ \ Si_s + HOH \longrightarrow
\end{array}
\quad
\begin{array}{c}
OH\ \ OH \\
\diagup\ \ \diagup \\
Si_s\ \ Si_s
\end{array}
$$

A surface siloxane group, for example, would react to form two hydroxyl silicon groups, known as silanols.

$$
\begin{array}{c}
O \\
\diagup\ \diagdown \\
Si_s\ \ Si_s + HOH \longrightarrow
\end{array}
\quad
\begin{array}{c}
OH\ \ OH \\
\diagup\ \ \diagup \\
Si_s\ \ Si_s
\end{array}
$$

The distribution of OH groups on the surface of fused silica [4] has probably been the most extensively investigated fundamental problem relating to the surface chemistry of siliceous glasses. This, in turn, is closely related to the nonreactive adsorption of molecular water on the glass surface.

II. THE SILANOL FUNCTIONALITY

A. Water Adsorption Isotherms and IR Spectra

G. J. Young [5], in 1957, reported the results of an extensive investigation into the surface properties of silica as a function of constitution, which laid the groundwork for present-day understanding of the subject. Samples of nonporous high-surface-area silica gel were placed in a vacuum system capable of reaching a pressure of 10^{-6} Torr, and studied by means of water-vapor adsorption and infrared absorption spectra. The measurements were made at room temperature after evacuating the samples for uniform periods of time at various temperatures ranging from 25 to 850°C (evacuation at elevated temperature is called "activation").

1. Room-Temperature Isotherms

Water-vapor-adsorption isotherms on the room-temperature-evacuated samples yielded isotherms intermediate between Type II and Type III*. Utilizing the value of $10.6A^2$ for the area occupied by a water molecule, surface areas for the silica-gel samples were estimated from the isotherms. The areas were obtained by using the BET method and also by selecting the point on the isotherm corresponding to the relative pressure where the monolayer value of water usually occurs in Type II isotherms. Although neither of these two methods are rigorous, due to the fact that the isotherms are partly Type III, the values obtained from the two methods were in good agreement with one another. The areas were found to be 1/4 to 1/8 of that found using the BET method with nitrogen adsorption, indicating that most of the silica-gel surface was hydrophobic. These isotherms were found to be completely reversible below a relative pressure of 0.5. At higher relative pressures hysteresis was found which, however, was due to effects resulting from packing of the particles.

2. Isotherms on Activated Samples

Water-vapor-adsorption isotherms obtained on samples which had been activated contained more Type III character than the room-temperature samples and indicated less adsorption. This trend increased with increasing temperature of activation throughout the entire range. When the samples that were heated up to 400°C were exposed to relative pressures less than 1.0 (so that no liquid condensation took place) and then evacuated, chemisorption was found to have taken place. This chemisorbed residue was found in each case to exactly equal the amount previously lost on activation. When adsorption isotherms were run on samples rehydrated in this manner, they were identical with those obtained before heating. The number of sites that were rehydrated was thus determined by first measuring the amount of H_2O that chemisorbed to a surface previously activated, and then measuring the amount of H_2O physisorbed to the surface after, as compared to before, chemisorption. When samples were activated above 400°, however, they could no longer be completely rehydrated. This effect increased with temperature of

---

*Type II isotherms are the common BET type which result from strong interaction followed by multilayer adsorption. Type III isotherms indicate limited or weak interaction between adsorbent and adsorbate.

activation until, at 850°, no chemisorption occurred on attempted re-hydration. There was still some physical adsorption on this type of surface, however.

3. The Silanol Group

The previously described behavior of the silica samples finds a ready explanation in terms of the presence and absence of silanol $(Si_sOH$, where Si_s is a surface silicon) groups on the surfaces. It is assumed that these hydroxyl groups are the active sites for physical adsorption of water molecules, and that they can be removed from the surface during activation by means of silanol condensation.

$$\begin{array}{cc} \overset{OH}{\overset{\diagup}{Si_s}} \; \overset{OH}{\overset{\diagup}{Si_s}} & \overset{O}{\overset{\diagup \diagdown}{Si_s \; Si_s}} \end{array}$$

The adsorption isotherms on room-temperature-evacuated samples are then interpreted as resulting from adsorption of water molecules to the surface in its state of full hydroxylation at 1/8 to 1/4 of the total number of surface sites. Vacuum heating of the sample in the range of room temperature to 180° accelerates removal of physisorbed H_2O molecules without affecting the total number of silanol sites. Activation between 180 and 400° removes silanols through condensation in increasing amounts with temperature. Upon readmission of water vapor, all siloxane sites formed from the previous silanol condensations are hydrolyzed and the silanol sites restored. Physisorption of water vapor then proceeds as for the unheated surface. Upon activation above 400° condensation of silanols increases further, and not all depleted sites can be restored through exposure to water vapor. This may be due to the possibility that rehydration through chemisorption must be preceded by physisorption at adjacent silanol sites. If dehydration above 400° has proceeded to the extent that many newly formed siloxane sites are not adjacent to any silanols, there may be no opportunity for rehydration of these sites by adsorbed water vapor.

4. Infrared Spectra

Infrared measurements on silica samples provided additional evidence for these proposed mechanisms. The samples were evacuated at the same series of temperatures as in the previous set of experiments. After each evacuation, the sample was cooled, exposed to water vapor at 0.4 relative pressure, and scanned to obtain the infrared adsorption spectra. Two peaks were observed which sharpened with increasing temperature of evacuation (Fig. 1). One peak was attributable to molecular water physisorbed to the surface and the other

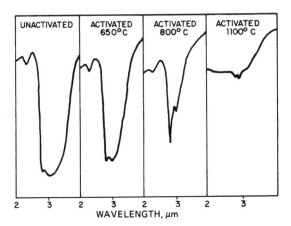

FIG. 1. Change in infrared spectra with sample heat treatment.
(Reprinted from Ref. 5, p. 79, by courtesy of Academic Press, Inc.).

to silanol groups on the surface. Both peaks decreased in intensity with
increasing temperature of evacuation above 400°, until at 1100° they
nearly disappeared.   These results confirmed the view that silanol
concentration on the surface decreased irreversibly with evacuation
above 400°, and that the physisorption of water vapor took place at
silanol sites (Fig. 2).

5.   Contamination Hypothesis

The unexplained hydrophobicity of 3/4 to 7/8 of the surface deserves
some comment at this point. Lacking proof to the contrary it could, and
perhaps should, be assumed that for some reason intrinsic to the structure and chemical nature of the silica surface, silanol groups cannot
form on more than one in four or one in eight sites. It has been suggested [6], however, that the prolonged evacuations and activations
of these samples in conventional vacuum systems which are not free
of organic vapors may have resulted in irreversible hydrophobic contamination of tne major portion of tne sample surface. The probability of this occurrence depends upon various details of the experimental
procedure (e. g., the nature of tne initial pumpdown and the location
and design of cold traps) wnich are rarely available in even the most
detailed write-ups. That this possibility should always be considered
in interpreting results of experiments in conventional vacuum systems
is demonstrated by a simple experiment [7]. A thin film of carbon is
deposited by evaporation in an ultraclean vacuum system onto a smooth,
polished, hydrophilic silica surface after removal from the vacuum system. The film is burned off in a hydrogen-oxygen flame. As the flame

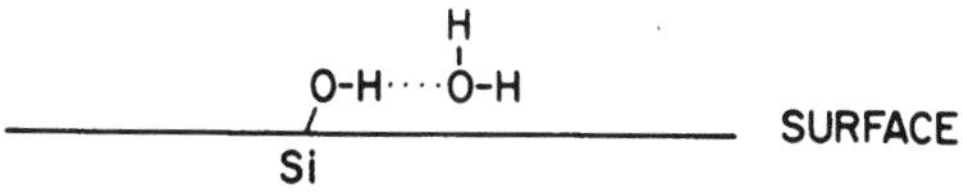

```
                      H
                      |
               O-H····O-H
    _____/_____ SURFACE
                 Si
```

FIG. 2. Water molecule hydrogen bonded to silanol.

temperature is increased to eliminate the glow due to the final remaining residue of carbon, a residue covering a sizeable fraction of the surface becomes increasingly difficult to remove. The residue clearly becomes embedded in the nearly molten silica surface where it is protected from oxidative removal. This embedded carbon is accompanied by a residual (non-zero) water contact angle.

B.   IR Spectra and Deuteration

1.   Freely Vibrating and Perturbed Silanols

Another advance in the understanding of the silica surface came with the publication by Kiselev et al. [8] of their (now classical) study of the infrared spectra of silica gels prior and subsequent to deuteration. Silica gels were heated in a vacuum at various temperatures and the IR spectra recorded. At 400°C activation, which Young had found to be the maximum temperature for reversible silanol condensation, the IR spectrum of the silica gel studied consisted of three distinct features. First, a narrow band was found at 3750 cm^{-1}, which persisted in spectra of all higher activation temperatures, including that at 800°. Second, a broad band was found extending over the region 3750-3000 cm^{-1}. Another wide band was found with a maximum at 3650 cm^{-1}. The broad 3750-3000 cm^{-1} band was sharpened by subtracting the 3650 cm^{-1} band from it, yielding a maximum of 3550 cm^{-1} for the 3750-3000 cm^{-1} band. This maximum will, henceforth, be used to designate the band. Continuous exposure of the silica gel to D$_2$O at room temperature resulted in complete elimination of the 3750 cm^{-1} and 3550 cm^{-1} bands, with no effect on the 3650 cm^{-1} peak (Fig. 3). The 3750 cm^{-1} peak was replaced by another at 2760 cm^{-1}, which resulted from the theoretically expected displacement of a freely vibrating OH peak when replaced by OD. The 3650 cm^{-1} peak was most persistent in silica samples with large globules, and underwent some deuteration when exposed to heavy water at elevated temperatures. All this constituted rather overwhelming evidence for the authors' conclusion that the peaks at 3750 cm^{-1} and 3550 cm^{-1} represented surface hydroxyl groups, while the 3650 cm^{-1} band resulted from internal hydroxyl. The sharp 3750 cm^{-1} peak was interpreted as representing "freely vibrating" surface hydroxyls, in which each individual hydroxyl is sufficiently isolated to avoid interaction with neighboring hydroxyls.

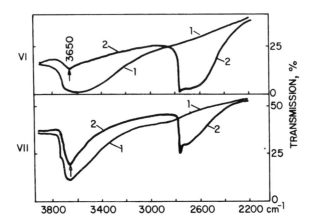

FIG. 3. Infrared spectra of silica gel samples VI (s = 340 m^2/g) and VII (39 m^2/g). (1) Before reaction with D$_2$O vapor; (2) after reaction with D$_2$O vapor at room temperature (followed by evacuation at 200°). [Reprinted from Ref. 8, p. 2257 by courtesy of the Chemical Society (London)].

The broad 3550 cm^{-1} peak was interpreted as consisting of "perturbed" surface hydroxyls, in which the individual hydroxyls were sufficiently close to one or more neighbors to undergo hydrogen bonding which resulted in a perturbation of the individual vibrations and a consequent broadening of the peak and shift to a lower frequency. When the surfaces were heated above 400°C, about 60 percent of the surface silanols remained, with the remaining 40 percent leaving the surface to form H$_2$O molecules, the product of the condensation of neighboring silanol groups. All of the remaining 60 percent of the original surface silanols were found by the authors to be freely vibrating hydroxyls.

2. Effect of Adsorption

For additional information on the properties of the silica surface, the authors allowed trimethylchlorosilane to interact with the surface under atmospheric conditions. This compound was undoubtedly chosen due to the widespread use of substituted silanes to alter the surface properties of glass fibers. They then evacuated the sample at 200°C for IR absorption measurements and deuteration. Another sample which was not treated with trimethylchlorosilane was similarly evacuated and used for comparison. It was found that the peak at 3750 cm^{-1} attributed to freely vibrating hydroxyls disappeared completely as a result of the trimethylchlorosilane treatment, while the 3550 cm^{-1} peak

assigned to mutually hydrogen-bonded surface hydroxyls remained vir-
tually unchanged. Simultaneously, peaks due to the methyl groups of
the substituted silane were introduced in the spectrum. From these
and other measurements on the sample, the authors concluded that
freely vibrating surface hydroxyls are active sites for the chemisorption of chlorotrimethylsilane, while mutually hydrogen-bonded surface
hydroxyl groups do not chemisorb the molecule. It was also suggested
that the freely vibrating surface hydroxyls are generally more reactive
to molecules possessing peripherilly concentrated electron density (in
effect, Lewis bases) than the mutually hydrogen bonded ones. Subsequently, Galkin, Kiselev, and Lygin [9] investigated the adsorption of
benzene, toluene, p-xylene, and mesitylene to these aerosil surfaces.
It was concluded that the freely vibrating surface hydroxyls were chiefly
responsible for the adsorption. It was also found that the magnitude of
the shift to lower frequency of the free hydroxyl groups upon adsorption
of these molecules was related to the heat of adsorption.

Armistead and Hockey [10] compared the reactions of chlorotri-
methyl-, dichlorodimethyl-, trichloromethyl-, and tetrachlorosilane
with aerosil silicas at 280°C by means of IR absorption spectra, deuteration, and postreaction chemical analyses of the aerosils for chlorine content. They found that the "free", or isolated, surface hydroxyl
groups reacted completely with the chloro- and dichlorosilanes. The
hydrogen bonded hydroxyls on the other hand, reacted solely, but in-
completely, with the tri- and tetrachlorosilanes. The reaction of the
mono- and dichlorosilanes with free hydroxyls was found by Cl analyses to be on a 1:1 (1 hydroxyl to 1 silane molecule) basis, except that
when heated to 500°C some bridging 2:1 reaction was found.

3. Pairing of Silanols

Peri and Hensley [11] investigated the distribution of OH groups
on the surface of silica gel with respect to their existence in pairs or
as isolated functional groups. Their use of the pair classification did
not differentiate between interacting or noninteracting pairs. For example, two silanols which are too far apart to undergo mutual hydrogen-bonding interaction, and are consequently designated as isolated,
or freely vibrating, hydroxyls with respect to their IR absorption properties, are nevertheless designated as paired if they are close enough
to interact simultaneously with one adsorbate molecule such as $SiCl_4$
or $AlCl_3$. The stoichiometry of the reaction of the surface hydroxyls
with $AlCl_3$ and $SiCl_4$ was used as the criterion of their existence in
pairs or in isolated form. It was assumed that a final stoichiometry
of two silanols to one adsorbate molecule was due to reaction of both
OH groups rather than reactivity of only half the groups. The authors
concluded that after activating the surface at 400°C more than 95%

of the surface silanols were paired using this definition, and even at 600° 85% were paired. They provided support for these experimental results by using a Monte Carlo method to randomly dehydrate the 100 face of beta crystobalite. The beta crystobalite face consists of a highly hydroxylated surface, with a total of approximately 8 hydroxyls per 100 A^2. Each Si atom on this postulated surface has two hydroxyls attached. The Monte Carlo random dehydrating procedure yields a resulting surface containing 4.56 hydroxyls per $100A^2$, all of which are either "geminal" or "vicinal" pairs. A geminal pair is one in which both hydroxyls are attached to the same silicon atom, while a vicinal pair consists of two hydroxyls on neighboring Si atoms. It is interesting to note that a geminal pair always consists of two freely vibrating hydroxyls which are not mutually perturbed and are consequently classified as isolated. This results from the tetrahedral bond angles of the Si atoms which orient the hydroxyls into positions too far apart to allow mutual hydrogen bonding.

Peri's finding that nearly all surface hydroxyls on silica gel exist as pairs even after drying at temperatures as high as 600°, as well as the corollaries that they must remain paired at any drying temperature, and that the surface of silica gel exclusively resembles the 100 face of crystobalite, was subsequently disputed by Hockey and coworkers [12]. They point out that "there exists a significant body of experimental evidence" that hydroxyl groups are present in substantial amounts both singly and in pairs on the silica surface. They refer to their own previously published works [13-15], in which experiments were performed on the reaction of various molecules with silica gels and powders. In these experiments, compounds such as $TiCl_4$, BCl_3, and $SiMe_2Cl_2$ were reacted with the silica surfaces and the silica substrate was ultimately analyzed for chlorine. Making the same assumption as Peri that a stoichiometry of 1:1 hydroxyl to multifunctional adsorbate molecule is proof of isolated hydroxyls (whereas 2:1 shows the existence of pairs), they concluded that on a "fully hydroxylated surface", i.e., one which has not been heated above room temperature after the hydroxylation procedure, 30% of a total of 4.6 hydroxyls per $100A^2$ are present singly while 70% are paired. A typical example of their procedure is the reaction of $SiMe_2Cl_2$ with a silica surface at 300°C. Analysis of the "solid reaction product" indicated the presence of chlorine equivalent to 1.3 Cl atoms per $100A^2$ of surface. Since the $SiMe_2Cl_2$ molecule has only two Cl functional groups, exclusive reaction with paired silanols as found by Peri would not leave any chlorine residue on the silica surface, whereas reaction with one silanol leaves one Cl on the surface. Furthermore, the amount of single OH's thus found (i.e., 1.3 per $100A^2$) corresponds to the number of "A" (single) sites found in IR work and other investigations. Hockey et al. explain Peri's contradictory results as being due to

his method of analyzing for HCl as a reaction product, which would include reaction of bulk hydroxyls emitted during subsequent heating of
the samples. Peri [11] on the other hand, disputes the presence of
bulk OH's in his samples.

4. Kinetics of Reaction with Substituted Silanes

Hair and Hertl [16] investigated the kinetics of reaction of gas-
eous, substituted silane molecules, mainly the chlorosilanes, at elevated temperatures, by following changes in IR spectra. They focused on the properties of the isolated (non-hydrogen bonded) silanols
by heating their samples to 800°C to drive off all adjacent hydroxyls.
The samples were then cooled to the reaction temperature of 150° to
300°, and the halosilanes admitted. The reaction kinetics were then
followed by observing the decrease in height of the peak at 3750 cm^{-1}
attributed to the isolated, freely vibrating hydroxyls. Reaction of one
halosilane with two surface hydroxyls leads to second-order kinetics
with respect to the hydroxyl groups, while reaction of one halosilane
with one hydroxyl, would lead to first-order kinetics. Reaction of
50% of the hydroxyls bimolecularly and 50% monomolecularly with
the halosilane, leads to a 1.5 order reaction. The authors found
that while chlorotrimethylsilane, which can only react with one hydroxyl group per molecule, always yielded a first-order reaction,
silane molecules with two or more chlorine groups yielded 1.4- to
1.7-order reactions. These kinetics, of course, require the conclusion that approximately one half the hydroxyls are isolated and one
half are paired. However, since all the OH groups were freely vibrating, they concluded that the paired groups are all geminal, i.e.,
two hydroxyls on one silicon. These geminal hydroxyls then, are
considered to be close enough to allow both to react with one substituted silane, but too far apart to hydrogen bond with each other.

5. Resolution of IR Peak at 3750 cm^{-1} into Three Components

An interesting by-product of these kinetic studies was evidence of
a very fast reaction which consumed about 15% of the OH groups at
close to zero time of the first- or second-order reaction which governed the reaction of approximately 70% of the remainder of the surface silanols. This was attributed to the well-known replacement of
an entire OH group on the surface by Cl. A search for the uniqueness
of those OH groups which were susceptible to this reaction did not provide a clear cut answer. However, in their search for different reac-
tivities among the isolated OH groups, Hair and Hertl [16] electron-
ically resolved the IR spectrum of these groups, and found a resolution
of the 3750 cm^{-1} peak in three peaks at 3743 cm^{-1}, 3747 cm^{-1}, and

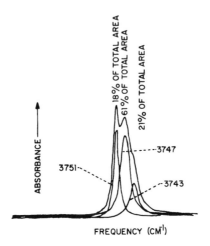

FIG. 4. Expanded spectrum of band due to freely vibrating hydroxyl group. The upper curve is the observed spectrum; the lower curves are the resolved bands and are composed of symmetrical Lorentzian shaped bands. [Reprinted from M. L. Hair and W. J. Hertl, J. Phys. Chem., 73, 2372 (1969) by courtesy of the American Chemical Society.]

3751 cm^{-1} (Fig. 4). They speculated that the 3747 cm^{-1} peak was due to single freely vibrating silanols, while the 3743 cm^{-1} and 3751 cm^{-1} peaks represented the geminal groups. This view received some support from experiments in which the 3751 cm^{-1} peak dropped faster than the other two upon reacting with chlorosilanes at 300°C. Some other adsorbates were found to react on the basis of two hydroxyls with one adsorbate molecule for the case of the 3751 cm^{-1} and 3743 cm^{-1} groups, while reacting one-for-one with the 3747 cm^{-1} group.

Uytterhoeven and coworkers [17] confirm the resolution of the 3750 cm^{-1} peak into three components, which they label HF (high frequency), MF (medium frequency), and LF (low frequency). Unlike Hair and Hertl [16] however, who assign HF and LF to one component, and MF to another, Uytterhoeven proposes that LF consists of one component, while HF and MF consist of another. This is supported strongly by deuteration, which replaces LF with its expected deuterated equivalent, while HF and MF collapse into one band. Uytterhoeven also differs with respect to assignment of the nature of the two components. He suggests that the LF component consists of isolated, freely vibrating hydroxyls slightly perturbed by lattice oxygens, while the HF-MF component does not consist of geminal hydroxyl groups, but rather of isolated hydroxyls further apart than

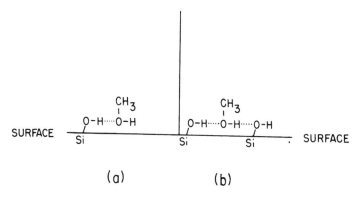

FIG. 5.   Hydrogen bonding to surface silanols.

geminals which are coupled through the electric fields of the vibrating protons.   Uytterhoeven points out, as evidence in support of this approach, that there is a correlation between energy of interaction, as measured by the magnitude of the IR frequency shift, and the increasing hydroxyl-to-adsorbent stoichiometric ratio of a group of physically adsorbed molecules of similar packing characteristics.   This implies, according to Uytterhoeven, that adsorbates capable of energetic interactions can interact at greater distances, thus providing an alternate mechanism for the increase in stoichiometric ratio.   It seems to us, however, that there is a special characteristic of strongly adsorbing molecules such as methanol and triethylamine, apart from mere ability to adsorb strongly, which may be responsible for this stoichiometric effect.   This is the inherent ability of any hydroxyl or amino group to act as an acid or base in hydrogen bonding, and furthermore to act as both.   The OH of a simple alcohol such as methanol will normally exhibit a strong preference for bonding to a surface silanol through the oxygen of the alcohol and the hydrogen of the silanol, rather than the oxygen of the silanol and the hydrogen of the alcohol (Fig. 5).   Now, when the alcohol acts as a base in this fashion, there is a shift of electronic charge in the molecule, towards the silanol hydrogen.   This leaves an increased positive charge on the alcoholic hydroxyl hydrogen, which then becomes more potent as the acid component of a hydrogen bond to be formed with the oxygen of a silanol.   Thus, when the oxygen of an alcohol acts normally in bonding to a silanol, it enhances the ability of the hydrogen of the alcohol to act abnormally and bond with a neighboring silanol, thus resulting in divalent physical bonding, peculiar only to hydrogen-bonding hydroxyl and amino functional groups.   If this is so, the surface silanols involved would have to be geminal, or equivalently close, paired, but freely vibrating hydroxyls.

C. Boria-Silica Surface

Hair and Hertl extended their studies of the kinetics of reaction of substituted silanes with silanol groups on silica surfaces to hydroxyl groups on combined boria-silica surfaces [18]. The surface preparation was similar to that used on the nominally pure silica surfaces, consisting of heating to 800°C to limit the hydroxyl presence to non-mutually-hydrogen-bonded groups. They found that while the boranol (B-OH) groups were more reactive than silanols, they also hydrolyzed more easily on exposure to vapor-phase molecular water. Another interesting feature of these surfaces was that the reactivity of the silanol groups was enhanced by the presence of the boron. Hair and Hertl also undertook a study of the vapor-phase reaction of PCl_3, BCl_3, CCl_4, Cl_2, $TiCl_4$, $SnCl_4$, $GeCl_4$, and $COCl_2$. In addition to the formation of Si_s - O - G bonds, where G is the atom to which Cl is originally attached, they found the well-known phenomenon of the replacement of an entire hydroxyl group on silica with a chlorine atom (SiOH + GCl → SiCl + GOH) to take place to a considerable extent. CCl_4 formed the SiCl species only, $GeCl_4$ yielded Si_sCl and Si_s-O-Ge in approximately equal amounts, $SnCl_4$ was less reactive than $GeCl_4$, yielding both types of surface species, $TiCl_4$ was very reactive, yielding a Si_sOTi type species, and PCl_3 was the least reactive, displacing about 25% of the hydroxyls, with the main product being the Si_sCl species.

III. NATURE OF SURFACE AT VARIOUS STAGES OF HYDRATION

The fundamental studies described in the previous section (II) focus upon the importance of silanols on the surface of siliceous glasses both in determining the physical and chemical reactivity of these surfaces with water vapor, and in determining surface reactivity with substituted silanes—the most versatile and effective class of surface-modifying agents for glass. While the studies described span the period from 1957 to the present, applications of these principles took place prior to and during the conduct of the fundamental investigations. This section describes wettability (and implicitly adhesion) phenomena which are best understood in the light of the fundamental surface properties previously described, while the following section (IV) describes the properties of fiber surfaces altered by chemisorption of substituted silane coupling agents.

A. Wettability by Non-Hydrogen-Bonding Liquid at Various Relative Humidities

The effect of water physisorbed to siliceous glass surfaces has probably been most dramatically demonstrated by Shafrin and Zisman [19] in their studies of the contact angle of methylene iodide, a non-hydrogen-bonding organic liquid, on glass surfaces in equilibrium with water vapor at various relative humidities. The angle of contact of a drop of liquid on a solid is a measure of the interaction of the liquid with the solid surface [20-28]. A large interaction tends to spread the drop over the surface to form a small or zero angle of contact, whereas a small interaction allows the surface tension of the liquid to maintain a more nearly spherical drop shape, resulting in a higher angle of contact with the surface. A clean, hard, solid surface, such as glass or metal, will interact strongly and yield a zero contact angle with all ordinary liquids [29, 30]. Shafrin and Zisman found a contact angle of 13° for methylene iodide on glass at a relative humidity (RH) of 1%, seemingly in violation of the rule that the contact angle of all liquids on glass should be zero. As the RH was increased, the contact angle increased, especially in the RH region above 45%. At RH 95%, the contact angle was 36°. This compares with a contact angle of 37° for methylene iodide on the surface of bulk water. The investigators point out that at high RH, a multilayer of adsorbed H_2O is formed which is thick enough to possess the properties of bulk water, and also mask the force field of the glass surface. As this thickness decreases with lower RH, the masking decreases and the contact angle is lowered. The residual contact angle of 13° at very low RH consequently implies residual adsorbed water. Adsorption isotherms of water on glass support this view. A composite isotherm of reports of various investigators in different pressure regions [31, 32] indicates a slow buildup of a single monolayer of adsorbed molecular water in the range of less than 1% to about 50% RH. At this point, multilayer formation starts and builds up rapidly to a thick film, with properties close to that of bulk water at 95% RH. Isotherms in the pressure region of 10^{-3} to 10^0 Torr of water [31], corresponding to RH between .005 and 5%, indicate small fractions of a monolayer remaining at very low RH. It is unclear, however, as to whether this residual moisture consists of molecular water alone. Assuming that all the adsorbed H_2O is removed, the question remains whether the remaining silanol groups, which can be regarded as chemisorbed water, continue to mask the high-energy surface, or whether the absence of physically adsorbed water is adequate to yield a high-energy surface.

B. Wettability by Non-Hydrogen-Bonding Liquid in Ultrahigh Vacuum

In an attempt to answer these questions, Schrader [33] investigated the contact angle of methylene iodide on smooth silica under ultrahigh vacuum conditions. A clean surface can be prepared in an ultrahigh vacuum apparatus by one of a number of methods [34, 35], such as vacuum heating [36-38], ion bombardment [39-42], vacuum evaporation [43] or vacuum cleavage [44, 45]. The surface thus prepared is then kept free of adsorbate by maintaining a suitably low pressure until the desired measurements are made. A pressure of 10^{-10} will safely maintain a surface free of all adsorbed ambients for a few hours. The measurement of contact angles on clean surfaces prepared in an ultrahigh vacuum is accomplished [33, 46] by admitting vapor into the evacuated sample chamber containing the clean solid surface, condensing the vapor to a liquid drop by means of a cold finger, and raising the sample magnetically to contact the drop. The vapor must be admitted in such a fashion that the <u>partial pressure</u> of possible contaminants is in the ultrahigh vacuum range. This can sometimes be accomplished by extensive degassing of the liquid before admitting its vapor to the sample chamber, such as in the measurement of the contact angle of oxygen-free water on metal surfaces [47]. The admission of methylene iodide free of any traces of water vapor is more difficult, however, due to the relative similarity of the vapor pressures of water and methylene iodide. The problem was solved by placing a porous silica gel with approximately 10 million square centimeters of surface in the sample chamber near the sample surface. This "getter" completely chemisorbed the remaining traces of moisture in the methylene iodide vapor and protected the sample from trace moisture throughout the contact-angle measurements. When the silica was vacuum baked in such a fashion as to yield a surface that was probably half silanol and half siloxane sites, a contact angle of $0°$ was obtained for methylene iodide. For an all-silanol surface, the contact angle was approximately $10°$. Admission of molecular water to the all-silanol surface caused a further increase in the contact angle of methylene iodide. Contact angles from $11°$ to $20°$ were attributed to surfaces of increasing fractional monolayer coverage of molecular H_2O, while an increase above $20°$ was attributed to the presence of a multilayer of adsorbed H_2O.

C. The High-Energy Glass Surface

The ability of the surface described as half silanol and half siloxane to yield a zero-degree contact angle with methylene iodide warrants further discussion. This surface was obtained after prolonged heating in ultrahigh vacuum culminating in 5 hours at $340°$. The most direct explanation of these results is that the siloxane groups are "high-energy" sites while silanol groups are not. A surface consisting of siloxane sites would spread methylene iodide with a zero contact angle, while a completely silanol-covered surface that is devoid of siloxane groups does not quite yield sufficient interaction to accomplish this. A surface composed of nearly half siloxanes, and the remainder of silanols, however, retains sufficient high-energy character to spread (zero contact angle) methylene iodide. Another possible approach is to postulate that freely vibrating, or isolated, silanols are the high-energy sites chiefly responsible for the interaction with methylene iodide. If reduction of the concentration of silanol sites on the surface reduces mutual hydroxyl interaction and creates a greater number of isolated silanols, a high energy surface could be created which would spread methylene iodide. In this case of course, complete removal of silanols, if possible, would yield a surface that does not necessarily spread the methylene iodide.

Now, while evidence such as that summarized in previous sections, based on IR and adsorption measurements, has been presented to indicate a greater physical adsorptivity on the part of "freely vibrating" than "perturbed" hydroxyls, none of the evidence indicates that the number of freely vibrating hydroxyls ever increases with a decrease in surface silanol concentration. Rather, the evidence indicates that a fully hydroxylated surface consists of a given number of freely vibrating and a given number of mutually interacting hydroxyl groups. As activation temperature is increased and hydroxyl groups are removed through silanol condensation, the number of perturbed groups decreases while the number of freely vibrating groups remains constant or decreases slightly. After all perturbed groups are removed, freely vibrating silanols become substantially depleted at sufficiently high temperatures. In other words, removal of mutually interacting silanols always occurs in pairs or clusters so that no new, freely vibrating hydroxyls are ever produced by this process. The ability of a surface consisting of half silanols and half siloxanes to spread methylene iodide while an all-silanol surface yields a $10°$ contact angle must therefore, on the basis of presently available evidence, be attributed to the high-energy character of the siloxane sites. The overall picture of the adsorption of water to glass and its effect on the "universal adhesive" properties of the glass is, therefore, as follows. The "high-energy" glass surface which possesses universal adhesion to all substances has the same chemical constitution as the bulk, consisting essentially of siloxane groups, possibly in strained form. A clean (devoid of organic contaminants) but "real" glass surface, defined as that which has been exposed to normal ambient molecules, is covered with a monolayer of silanol groups, resulting from exposure to atmospheric moisture. This surface is no longer a theoretical high-energy solid surface in the usual sense, but, nevertheless, maintains a sufficiently high surface free energy to spread non-hydrogen-bonding liquids with surface tension less than 50 dynes per centimeter. Its activity with respect to hydrogen bonding materials will, of course, be much greater. Upon adsorbing a monolayer of water on these silanol groups, the overall surface activity of the glass will again be reduced. Adsorption of a multilayer of water will further reduce the surface energy, especially that component capable of short-range-dispersion interaction, and tend to give the surface water-like properties. Further thickening of this layer will result in the formation of liquid water on the glass surface resulting in lubrication of the surface and complete loss of meaningful adhesion to other materials.

D. Organic Contamination from Atmosphere

Another factor affecting the surface characteristics of glass under practical conditions is its tendency to pick up contamination consisting of large molecules of oily or greasy materials dispersed in the atmosphere. These substances are weakly adsorbed to the glass surfaces, reduce their surface energy and general adhesive capacity, and impart to them a partially hydrophobic character. The latter characteristic is often manifested by the formation of irregularly shaped puddles which stick to the glass surfaces. These differ from uncontaminated glass surfaces which spread water in a thin, transparent film, and from completely hydrophobic surfaces such as paraffin on which the water will curl up with a high contact angle and easily roll off the surface. The puddles on the contaminated surfaces often have undesirable effects such as decreasing visibility on the windshields of moving vehicles. On glass fibers this phenomenon could promote retention of water by inhibiting run-off of smooth films or rolling drops.

IV. GLASS-RESIN INTERFACE

A. Susceptibility to Water Penetration

Glass fibers are the major reinforcing material utilized in glass-reinforced plastics (GRP), which provides the highest strength to density of any material known. GRP utilizes the high tensile strength of glass while eliminating the brittleness and poor shock resistance of all-glass structures. The ability of GRP to act as a single material possessing the assets of both its components depends however, on good adhesion between the matrix and reinforcement. Poor adhesion can manifest itself as decreased composite tensile or compressive strength, or in lack of ability to maintain its normal strength on exposure to water under adverse conditions. The hydrophilic nature of the glass surface, resulting from the tendency of H_2O to form strong hydrogen bonds with the silanol groups of the silica network, and increased further by the presence of alkali metal and other ions, results in the strong tendency of water to enter the interface between glass fiber and resin, causing debonding and premature failure of the material. This instability of the glass-plastic interface on long-term exposure to water becomes readily apparent under conditions of accelerated exposure to moisture such as the commonly used 2-hour "water-boil" test. In this test, a 2 x 4 x 1/4-inch piece is cut from the reinforced plastic and boiled in water for two hours. Its flexural strength after this exposure compared to that of a "dry" piece gives the "wet strength retention" of the particular laminate. A 40% loss in flexural strength for polyester composites, and 25% for epoxies, is not at all uncommon after the 2-hour water boil. If the liquid water is marked with a dye or ink, it can be seen to enter the glass-plastic interface through the edges and then travel along the length of the fiber (at the interface) [48], debonding the plastic from the fiber along the way. In the event the liquid water is prevented from entering the interface through the edges, it will permeate the GRP in the vapor phase and ultimately condense as a liquid in the fiber-resin interface, once again debonding resin from glass.

B. Coupling Agents and Chemical Bonding Theory

An obvious possibility for solving this problem would be to apply a tenacious hydrophobic coating to the glass fiber before laminate preparation so as to provide continuing water repellency. One such example would be a substance such as trimethylchlorosilane, which can chemisorb to the glass surface via siloxane bond formation and create

$$Si_sOH + SiCl(CH_3)_3 \rightarrow Si_sOSi(CH_3)_3 + HCl$$

a new hydrophobic surface of bound CH_3 groups. The problem here, though, is that the CH_3 groups have low interaction with the resins as well as with water, thereby acting as a parting agent and ruining even the dry strength of the laminate [51]. To eliminate this difficulty, R. K. Witt [52] in a previously classified report submitted to the Navy in 1947, originally proposed to improve interfacial adhesion in glass-fiber-reinforced plastics by treating the fiber surface with a compound that would form a bridge of chemical bonds between fiber reinforcement and resin matrix. The concept called for a single molecule with two different functional groups, one of which would react with the glass and the other with the resin. In the 1950s a few such surface finishes were successfully introduced and in the 1960s many compounds were commercially available which improved the properties of a variety of resin-glass systems [62-68]. The molecular-bridge concept which stimulated these developments has become known as the "chemical bonding theory" [69], and the finishes applied to the fiber surface accordingly called "coupling agents." An example of the proposed mechanism of action of coupling agents is the use of vinyltriethoxysilane [70] as a fiber-surface finish in glass-reinforced polyester resins. According to the chemical bonding theory,

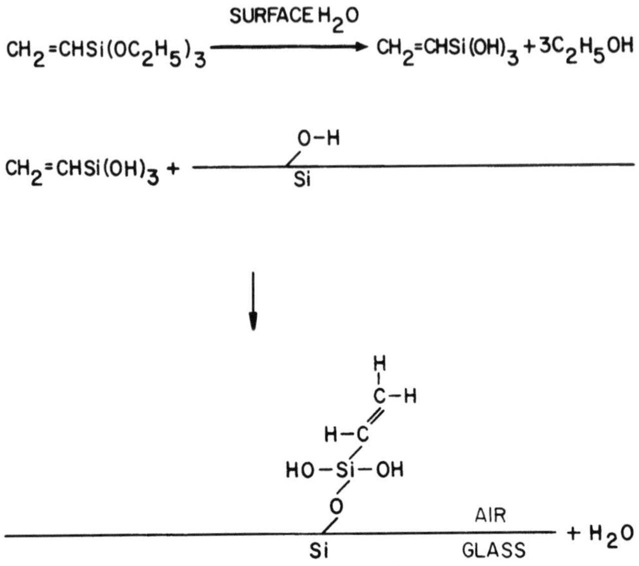

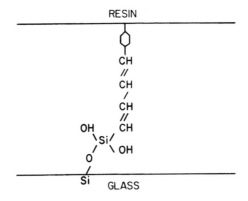

$$CH_2=CHSi(OC_2H_5)_3 \xrightarrow{\text{SURFACE } H_2O} CH_2=CHSi(OH)_3 + 3C_2H_5OH$$

$$CH_2=CHSi(OH)_3 + \underset{\begin{smallmatrix} | \\ Si \end{smallmatrix}}{\overset{O-H}{\diagup}}$$

FIG. 6. Chemical bonding of vinyltriethoxysilane to glass.

FIG. 7. Chemical bonding to polyester resin of glass treated with vinyl functional silane.

this compound would chemically react with glass (in two steps) via the ethoxysilane functional group to form the treated fiber surface (Fig. 6). Upon adding polyester resin to the glass fibers, the vinyl group copolymerizes with styrene unsaturation in the resin to complete the bridge of chemical bonds (Fig. 7). Another example is the use of gamma-aminopropyltriethoxysilane, $NH_2(CH_2)_3Si(OEt)_3$, as a coupling agent

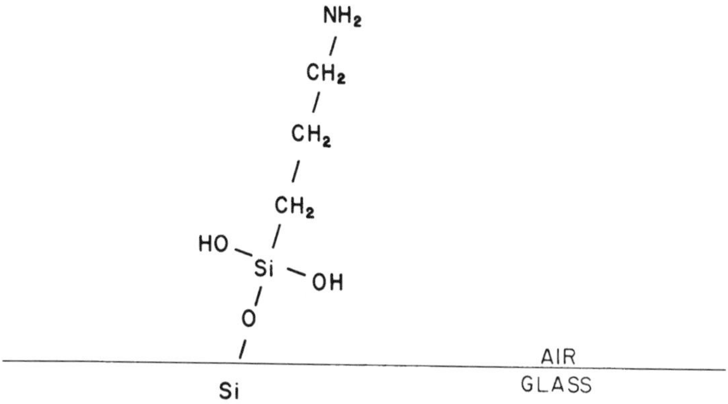

NH₂
|
CH₂
|
CH₂
|
CH₂
|
HO—Si—OH
|
O
|
Si

_____ AIR
                                            GLASS

FIG. 8.   Chemical bonding of gamma-aminopropyltriethoxysilane to glass.

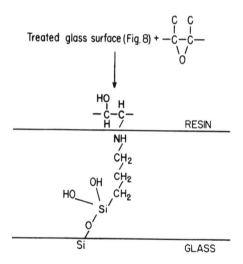

FIG. 9.   Chemical bonding to epoxy resin of glass treated with amino functional silane.

for glass-fiber-reinforced epoxy resins.   The ethoxysilane reacts with the glass surface in the same fashion as the vinyl compound of the previous example to form the treated glass surface (Fig. 8).

The amino group then reacts with epoxy functionality to complete the chemical bridge (Fig. 9).   While the chemical bonding theory has been the only workable hypothesis for the development of new finishes, or coupling agents, for various systems, its success has been qualified, and research in the area has been semiempirical.   Probably the best way of summing up the situation is that a proposed coupling agent for a fiber-resin system which does not meet the requirements of the chemical bonding theory will not work, and a proposed coupling agent which does meet these requirements may or may not work. The obvious conclusion from this is that the requirements of the theory are necessary but not sufficient to the functioning of coupling agents. Investigation of the physical and chemical nature of the coupling-agent-treated glass surface has shown that the coupling agent is present on the surface in a far more complicated configuration than had been expected from the simple picture of a monomeric chemical bridge. The existence of factors additional to those considered by the original chemical bonding theory is consequently not surprising.

C. Nature of Coupling Agent on Fiber Surface

1. Multilayer Formation

Sterman and Bradley [56], using a technique developed by D. E. Bradley for replication of fibers [71], first applied the electron microscope to an investigation of the amount and state of aggregation of various silane coupling-agent films applied to E-glass fibers. The first and most striking feature of their findings was that the commercial method of applying coupling agents resulted in deposition of a thick (in terms of molecular dimensions) nonuniform layer of the material, which tended to form agglomerates in the cavities between the fibers. Furthermore, upon extracting the treated fibers in a Soxhlet for 4 hours, about 80% of the deposited coupling agent was removed, with the remainder present in the form of islands. The technique was not sensitive enough to determine whether the space between the islands was bare glass or contained an ultra-thin film on the order of a monolayer. A comparison of the material present in the islands as estimated by electron microscopy with the total amount present (on the extracted surface) as determined by chemical analysis did not necessarily require the postulating of coupling agent in between the islands to obtain a material balance. In short, the nature of the surface layer of coupling agent as seen by the electron microscope was far different than the simple picture of a monolayer of neatly oriented molecules required by the chemical bonding theory. The authors concluded that the film consisted mainly of easily removable outer layers but with a "tightly bonded polymer close to the glass." To obtain optimum performance from the coupling agents, the authors found it necessary to apply from 8 to 70 monolayer equivalents to the glass surface. They explained the need for high loadings on the basis of the uneven distribution of coupling agent on application.

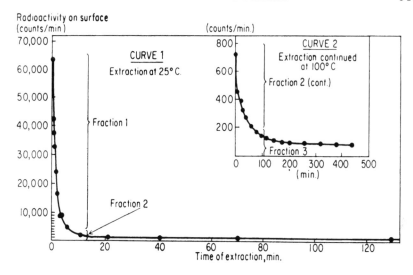

FIG. 10. Amount of radioactive APS remaining on polished Pyrex surface after extraction with water. (Reprinted from Ref. 72, p. 196, by courtesy of McGraw-Hill, Inc.)

2. Heterogeneous Nature of Multilayer

Schrader, Lerner, and D'Oria [72-73] utilized radioisotope techniques in combination with electron microscopy in an attempt to further examine the nature of a coupling-agent film on glass. Of special interest was the question of the possible existence of a monolayer, which could not be detected with the electron microscope, in between the islands observed by Sterman and Bradley after Soxhlet extraction. The coupling agent chosen for investigation was gamma-aminopropyl-triethoxysilane (APS) with a C-14 label placed in the hydrocarbon chain alpha to the amino group. The coupling agent was deposited on the surface of polished Pyrex blocks from benzene solution, and the surface of the blocks, excluding the edges, was then counted for radioactivity. The blocks were then extracted in water at room temperature and the radioactivity on the block determined as a function of time of extraction. Following this the blocks were placed in boiling water and the residual radioactivity determined periodically. It was found that the deposit of APS could be regarded as consisting of three fractions with respect to tenacity when subjected to extraction procedures (Fig. 10). The major fraction, called Fraction One, which was as much as 98% of the total, depending on the amount deposited, consisted of the hydrolyzate of the silane (ethoxy groups replaced by hydroxyls) physically adsorbed to the surface. It could consist of as much as 270 monolayer equivalents, and although insoluble in benzene it was rapidly removed by the cold water rinse. Fraction Two was a chemisorbed polymer of the coupling agent, consisting of about 10 monolayer equivalents. This fraction required about 3 to 4 hours extraction in boiling water for essentially complete removal. Electron microscope pictures of this fraction taken in parallel experiments after a one-hour boiling-water extraction showed the presence of islands. Assuming that the one-hour water boil is roughly equivalent to a four hour Soxhlet extraction, these islands can be identified with those observed by Sterman and Bradley, supporting their findings of a tightly bound polymeric layer. Now, upon complete removal of this Fraction Two, defined by the complete tapering off of the curve of radioactivity versus time obtained during the boiling-water extraction, Schrader et al. [72] found that the electron microscope now indicated a completely bare surface, the islands having all disappeared. However, a radioisotope count of this apparently bare surface indicated that the equivalent of a monolayer (or substantial fraction thereof) of the previously undetected coupling agent still remained. This residue was called Fraction Three. It was apparently more tenaciously held to the surface than Fraction Two, the bulk of the chemisorbed polymer, and the authors speculated that this resulted from multiple bonding of the molecules (perhaps as monomers, perhaps as dimers or trimers) to the glass surface.

3. Hydrolytic-Desorption Kinetics

Data which is presented in a subsequent section of this chapter indicate that Fraction Two, the chemisorbed coupling-agent "multilayer," is vital to coupling-agent function. The kinetics of desorption [73, 74] of the chemisorbed coupling agent from bare glass in boiling water yields significant information on its structure. If the coupling agent were a cross-linked polymer, assuming random-scission, there would be an induction period for the desorption process, since no desorption can take place until hydrolyzed bonds completely sever a segment from the matrix. For the case of the adsorbed polymer chains with little or no lateral linking (Fig. 11), however, the rate of reaction is greatest initially, decreasing as the reaction proceeds. The simplest case of this latter type would be the first-order reaction resulting if hydrolysis can take place only at the bond joining each polymer chain to the surface. The rate is then proportional to the amount on the surface at any given time, and a plot of the logarithm of the amount on the surface versus time will yield a straight line.

If hydrolysis can take place randomly anywhere in the polymer chain, the following considerations prevail:

$$-\frac{dc_1}{dt} = kc_1, \quad -\frac{dc_2}{dt} = 2kc_2, \quad -\frac{dc_3}{dt} = 3kc_3, \cdots, \quad -\frac{dc_n}{dt} = nkc_n, \cdots \quad (1)$$

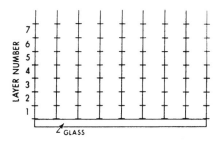

FIG. 11.   Separately adsorbed polymer chains.   (Reprinted from Ref. 74, p. 286, by courtesy of John Wiley and Sons, Inc.)

$$-\frac{dc_m}{dt} = -\frac{dc_1}{dt} - \frac{dc_2}{dt} - \cdots - \frac{dc_n}{dt} = kc_1 + 2kc_2 + \cdots + nkc_n \qquad (2)$$

$$= \sum_{i=1}^{i=n} \left(-\frac{dc_i}{dt}\right) = k \sum_{i=1}^{i=n} ic_i \qquad (3)$$

where c_1 is the concentration of monomeric units attached to the glass; C_2 that of monomeric units one layer removed from the glass, etc; k the first order velocity constant for scission of any layer from the one below it; t the time; n the thickness of the layer in monomeric units; and c_m the total concentration of monomeric units.   Integrating equation (1), we have

$$\ln c_1 = -kt - c \qquad (4)$$

where c is the integration constant.   At t = 0, $-c = \ln c_1^0$, where c_1^0 is the initial c_1 concentration.   Therefore, $\ln c_1 = -kt + \ln c_1^0$, and we have

$$\ln \frac{c_1}{c_1^0} = -kt \qquad (5)$$

Likewise, $\ln (c_2/c_2^0) = -2 kt$, and since $c_1^0 = c_2^0 = c_3^0 = c_n^0$, we have $\ln (c_2/c_1^0) = -2 kt$

$$\ln \frac{c_3}{c_1^0} = -3 kt, \cdots, \ln \frac{c_n}{c_1^0} = -n kt \qquad (6)$$

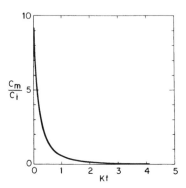

FIG. 12. Hydrolytic desorption of segments of separately adsorbed polymer chains: theoretical curve. (Reprinted from Ref. 78, p. 116, by courtesy of Academic Press, Inc.)

or,

$$c_1 = c_1{}^0 \, e^{-kt}, \quad c_2 = c_1^\circ \, e^{-2kt}, \cdots, \quad c_n = c_1^\circ \, e^{-nkt} \tag{7}$$

$$c_m = c_1 + c_2 + \cdots + c_n \tag{8}$$

$$\frac{c_m}{c_1{}^0} = e^{-kt} + e^{-2kt} + \cdots + e^{-nkt} \tag{9}$$

$$\frac{c_m}{c_1^\circ} = \frac{e^{-nkt} - 1}{1 - e^{kt}} \tag{10}$$

A typical curve generated by this equation for a polymer 10-monomeric-units thick is given in Figure 12 [78]. This may be compared with the experimental curve in Figure 13. The similarity between the general shape of the experimental curve and that predicted by the model is strong evidence of the existence of a rod-like chemisorbed structure with little or no cross-linking between the rods.

Of special interest are the final and initial stages of the reaction. It is obvious that in the final stage the reaction approaches first order,

$$\frac{-dc_m}{dt} = kc_m$$

since the average length of the rods is continually reduced as desorption proceeds. At commencement of desorption,

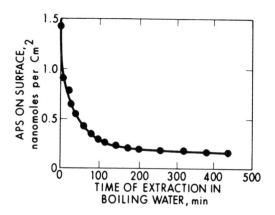

FIG. 13. Hydrolytic desorption of chemisorbed APS. (Reprinted from Ref. 74, p. 283, by courtesy of John Wiley and Sons, Inc.)

$$c_1^0 = c_2^0 = c_3^0 = \cdots = c_n^0; \quad c_m^0 = c_1^0 + c_2^0 + \cdots + c_n^0 = nc_1^0$$

$$\frac{-dc_m^0}{dt} = kc_1^0(1 + 2 + 3 + \cdots + n) = \frac{n(n+1)}{2} kc_1^0 \tag{11}$$

where the superscript ^{0} refers to initial conditions. Now,

$$c_m^0 = nc_1^0$$

$$\frac{-dc_m^0}{dt} = \frac{n+1}{2} kc_m^0 \tag{12}$$

The maximum ratio of the initial first-order velocity constant to the final one is then given by

$$\frac{(dc_m^0/dt)/c_m^0}{(-dc_m/dt)/c_m} = \frac{(n+1)/2}{k} k = \frac{n+1}{2} \tag{13}$$

or

$$\frac{-d \ln c_m^0/dt}{-d \ln c_m/dt} = \frac{n+1}{2} \tag{14}$$

A ratio greater than $(n + 1)/2$ may therefore be taken as evidence of heterogeneity of the surface-adsorbed polymer with respect to desorption, i. e., that the last portion to desorb is more strongly bound or intrinsically more difficult to desorb than the original portion. Of course, the rule is not necessarily conclusive, since the physical nature of the real multilayer must agree with that of the model in all important aspects. For example, a multilayer with many holes in it would have a real value of n (thickness of the multilayer) which is greater than the value calculated from the amount adsorbed per unit of surface area. The sharpness of taper of the curve would therefore be greater than expected due to this factor, rather than to more strongly bound residue.

It can be seen that the curves in Figures 12 and 13 parallel each other fairly closely until the final stage which represents the desorption of less than a monolayer equivalent. At this point, the upper curve tapers off more sharply. Using the criterion of Equation (13), it can be seen that the ratio of the initial to final slope is far greater than $1/2$ $(n + 1)$, yielding evidence that the last portion to desorb is more strongly bound, or intrinsically more difficult to desorb, than the original portion. Equivalently, it may be observed that whereas the theoretical curve effectively reaches zero in a reasonable time period, the experimental curve runs nearly parallel to the abscissa, incidating a strongly adsorbed submonolayer species. All in all, therefore, the behavior observed strongly suggests a model for the surface structure of separately adsorbed polymer chains, as in Figure 11, which are depleted from the interface in the presence of boiling water as a result of hydrolysis of siloxane bonds linking the monomeric segments. Also, the link, or links, between the bottom segment and the glass surface, are probably not equivalent to the intersegmental links, and are hydrolyzed with greater difficulty.

D. Coupling Agent at Glass-Resin Interface

1. Identity of Coupling-Agent Fraction Responsible for Wet-Strength Retention

The presence of coupling agents on the surface of glass in the form of a heterogeneous film raises the question as to which of the components of this film are responsible for its effectiveness in protecting the resin-glass interface from degradation by exposure to water; what is the minimum, in terms of the surface chemical structure and amount of a component or components, necessary to provide this protection; and, what combination of conditions yields maximum protection. Schrader, Lerner, and D'Oria [72] and then Schrader and Block [74] determined the efficacy of each of the fractions of the heterogeneous film by separately measuring the effectiveness of each in protecting glass-epoxy adhesive joints from the effect of moisture. Pyrex glass-block surfaces containing selected components of the heterogeneous

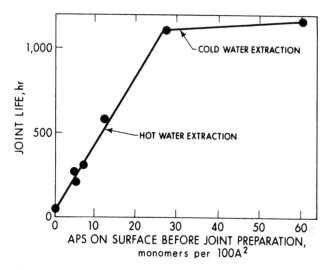

FIG. 14. Joint life as function of APS applied to surface by deposition-extraction method. (Reprinted from Ref. 74, p. 283, by courtesy of John Wiley and Sons, Inc.)

film, in varied amounts, were prepared by varying the nature of extraction with solvent following deposition of the coupling agent. The treated blocks were then crossed, in pairs, and cemented together with epoxy resin. Each glass-epoxy-glass adhesive joint was then immersed under a 50 lb load in hot water, and the time for failure of the joint ("joint life") automatically recorded. After a few preliminary results were reported, Schrader and Block [74] investigated the entire range of coupling agent coverage utilizing radioactive APS on all Pyrex blocks which were used to form the adhesive joints. This allowed direct determination of the residual APS on the block prior to preparing the joint. It was found that joint lives of equal duration were obtained with or without the presence of Fraction One (physically adsorbed hydrolyzate which is removed by cold water) in the heterogeneous coupling-agent film. In fact, too large an excess of Fraction One sometimes caused bond deterioration. Maximum joint lives were obtained with maximum amounts of Fractions Two and Three (chemisorbed polymeric APS). With depletion of Fraction Two by extraction with boiling water before preparing the adhesive joint, there was a linear decrease in joint life (Fig. 14).

2. Effect of Method of Application on Bond Life

In another set of experiments [74], the glass blocks were immersed in a very dilute benzene solution of radioactive APS, then

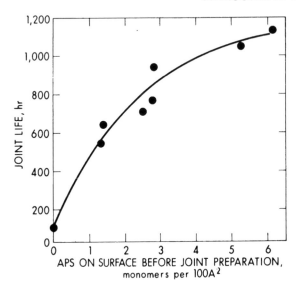

FIG. 15. Joint life as function of APS applied to surface by adsorption method. (Reprinted from Ref. 74, p. 284, by courtesy of John Wiley and Sons, Inc.)

withdrawn, drained, and room-temperature cured. The amount of coupling agent adsorbed on the surface was regulated by the time of immersion. When applied under these conditions, the near-maximum joint life of 1100 hours was obtained with two monolayer equivalents of coupling agent (Fig. 15) whereas approximately 8 monolayer equivalents were required to obtain this joint life when the coupling agent was applied by the deposition-evaporation technique followed by cure at 110°C.

3. Mechanism of Hydrolytic Failure

The question of the mechanism of hydrolytic failure of a glass—coupling-agent—resin adhesive joint is entirely different, of course, from that of the mechanism of extraction of coupling agent from bare glass in hot or boiling water. First, the more complicated interface has additional components which may become a weak link, such as, for example, the coupling-agent-to-resin bond. Second, the nature of the coupling agent may change as a result of its interaction with the resin. Third, and perhaps most important, is the protection which the resin, in combination with the coupling agent, gives with respect to the entrance of water into the interfacial region. A resin—coupling-agent—glass adhesive joint can take weeks or months to fail

in hot water, even under load, as compared to hours required to remove nearly all the coupling agent from bare glass immersed in hot
water. Interestingly, James, Norman, and Stone [75] have found
evidence that, with coupling agent present in a laminate, water migrates to the interface as a vapor rather than as a liquid. It is clear
then, that the mechanism for failure at the resin-coupling-agent-glass
interface must be separately determined.

The possible loci for homogeneous debonding in the interface region may be listed as follows:

1. In the resin near the interface

2. At the resin—coupling-agent subinterface

3. In the coupling-agent structure at the interface

4. At the coupling-agent—glass subinterface

5. In the glass near the interface.

It is obvious that knowledge of the disposition of a radioactively
labeled coupling agent after debonding of a single epoxy—coupling-
agent—glass joint would narrow down this list of possibilities considerably. For example, if all the coupling agent remained on the
glass, then either (1) or (2) of the above possibilities is true. If
coupling agent remained on glass and resin, the implication is that
(3) is the mechanism, if debonding is homogeneous. If all the coupling agent remained on the resin, (4), or (5), would be indicated. In
nearly all the bond life tests performed by Schrader and Block [74]
the failure appeared to be cleanly interfacial so that the resulting
glass surface, and often the debonded resin, could be examined for
radioactivity. The results for the amount of radioactivity remaining on glass which was debonded from glass—coupling-agent—epoxy-
resin adhesive joints are given in Figure 16, where the amount of
APS on a glass-block surface after failure of a joint is plotted versus the amount that was on that surface prior to making up the joint.
It can be seen that throughout this entire range of surface coverage
by adsorption (approximately two monolayer equivalents based on one
molecule per $33A^2$), half the coupling agent remains on the surface
after joint failure, and half has been removed.

There are three aspects to this observed disposition of the labeled coupling agents. First, the fact that radioactivity is found on
both sides of the interface implies that failure took place in the coupling-agent structure itself. Second, the fact that the fraction remaining on the glass surface is reproducible over many samples provides
strong evidence that this is indeed so, since a vertically zig-zag tear-
ing mechanism would result in wide variations from sample to sample.
Third, the fact that the reproducible fraction remaining on the glass
is equal to 0.5 suggests interesting speculation regarding the molecular

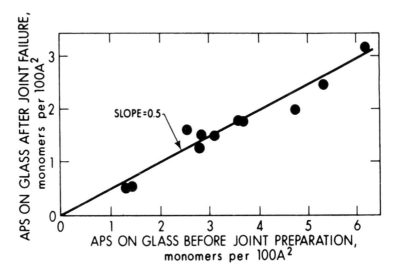

FIG. 16. Location of radioactive APS after joint failure: residue on glass (Reprinted from Ref. 74, p. 285, by courtesy of John Wiley and Sons, Inc.)

nature of the debonding. It is first assumed that the APS molecules adsorb to the glass surface as dimers, with one end of each dimer chemisorbed to the glass surface (Fig. 17) so that a double-monolayer equivalent is actually a monolayer of dimers, while a monolayer equivalent is one half a monolayer of dimers, etc. It is further assumed that the resin reacts only with the amino group of the molecule (half-dimer) that is not chemisorbed directly to the glass surface. This latter assumption is plausible since the amino group of the half-dimer which is attached to the glass is probably strongly adsorbed to the surface and consequently may be unavailable for reaction. Failure of the joint then results from hydrolysis of the siloxane linkage at the mid-point of each dimer which has reacted with both the resin and glass. Half of each dimer remains on the glass and the other half is removed, resulting in the observed constant 50% residue of radioactivity after joint failure. It is to be emphasized, however, that the hypothesized adsorption of APS to the glass surface in the form of dimers is peculiar not only to the process of adsorption from solution (as opposed to deposition followed by solvent evaporation), but also to the specific solvent (benzene) and concentration used in this set of experiments. Under different conditions, adsorption of APS quite possibly may take place as oligomers of mixed sizes. In those cases, it would be expected that hydrolysis of siloxane bonds of the oligomer would be the mechanism of hydrolytic failure. This is supported by the fact that radioactivity is found on both sides of the failure plane in all cases.

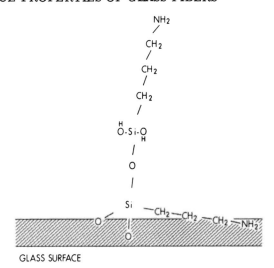

FIG. 17.   APS dimer chemisorbed to glass.   (Reprinted from Ref.
74, p. 289, by courtesy of John Wiley and Sons, Inc.)

E. Chemical Bonding to Resin

For the case of the joints prepared from glass treated by the adsorption method, only 0. 1 of the APS originally applied was retained
by the resin after joint failure. Since 0. 5 remained on the glass, the
results suggest that only about 0. 2, on the average, of the dimeric
coupling-agent molecules adsorbed to the glass have also formed chemical bonds with the resin. The coupling dimers hydrolyze (scission at
the midpoint) at the same rate as those adsorbed to the glass only, so
that when the joint fails all (or nearly all) the dimers, coupled or un-
coupled, have undergone scission. Half the radioactivity remains on
the glass, since each dimer has left a hydrolyzed monomer. Two
tenths of the dimers leave a hydrolyzed monomer attached to the resin,
accounting for one tenth of the original radioactivity, while the remaining four tenths goes into solution. The chemical coupling of only two
tenths of the dimers to the resin is consistent with molecular models,
which show the limited availability on the epoxy resin of sites reactive to the amino group of the coupling agent.

While chemical bond formation between silane functional groups
and the surface of glass is fairly well established, evidence that the
observed bonding of the coupling agent to the resin is indeed chemical in nature has been more elusive. Sterman and Marsden [76] determined the extractibility with benzene and trichloroethylene respectively, of polystyrene and polyethylene from composites with unfinished

E glass and E glass finished with 3-methacryloxypropyltrimethoxysilane coupling agent. For the case of both polymers, residual carbon analysis indicated that a thin layer of polymer was retained on the surface through long periods of extraction, as a result of the presence of the coupling agent. Johannson et al. [77] utilized carbon-14-labeled methylmethacrylatestyrene copolymer to determine the extractibility with tetrahydrofuran of that resin from its composite with E glass in the presence and absence of coupling agent. They found that the presence of coupling agent resulted in retention of a sizeable residue of resin after the extraction procedure. The results of both of these investigations favor the theory of chemical bond formation between resin and coupling agent.

V. SUMMARY

A clean, freshly formed glass surface consists mainly of exposed siloxane linkages which have the characteristic properties of "high-energy" surfaces, i. e. , an ability to undergo relatively strong physical interactions with all other substances. One easily observable manifestation of this property is the ability to spread all liquids (except mercury) at room temperature in opposition to the retracting force of their surface tensions. When exposed to atmospheric water vapor or liquid water, the siloxane groups react to form silanols. These surface hydroxyl groups reduce the ability of the silica network to undergo strong physical interaction with all substances, but increase the capability for hydrogen bonding and certain specific chemical reactions. An immediate result of the hydrogen-bonding capability is the formation of an adsorbed layer of water whose thickness increases with relative humidity. This results in further masking of the high energy of the surface with respect to non-hydrogen-bonding interactions, until with sufficiently thick adsorbed layers, the surface resembles that of bulk liquid water.

Removal of adsorbed molecular water through vacuum treatment reveals that the silanol groups are present in two major categories, those that are relatively isolated and those that are paired at sufficiently close distances to mutually hydrogen bond. The isolated silanols tend to be the more reactive species. However, reaction with multivalent atoms of ambient molecules such as silicon in substituted silanes often depends more upon suitable intrapair distance with respect to the stereochemical requirements of the reacting silicon than the degree of intrapair hydrogen bonding. Boria on the glass surface tends to form hydroxyl groups more rapidly than the silica network. The silicon-oxygen-boron linkage tends to hydrolyze more rapidly than the siloxane linkage.

The ability of a substituted silane to react with more than one surface silanol has resulted in the widespread use of these compounds as
agents to modify the characteristics of the fiber surface. These agents
are organosilicon compounds which consist of a silicon atom attached
to three functional groups capable of promoting reaction with the glass
surface or polymerizing with a fellow molecule, and a fourth functional
group which gives the glass surface its new interfacial property. If
this functional group is hydrophobic the fiber becomes water repellent.
If it is capable of copolymerizing with a resin, it is known as a coupling agent which provides improved adhesion at the interface in glass-
fiber-reinforced plastics. Radioisotopic and electron microscopic
studies have indicated that these coupling agents chemically bond to
the glass surface in the form of non-cross-linked oligomers. A minor portion of these oligomers chemically bond with resin during
formation of the reinforced plastics. The resulting bridge of chemical bonds between resin and plastic forms an interface with dramatically improved resistance to debonding by water penetration, and often
with increased dry strength as well. When the glass—coupling-agent—
resin interface does fail under long-term exposure to hot water (a
widely used, accelerated moisture exposure test), failure is caused
by hydrolysis of bonds within the coupling-agent oligomer. The very
high resistance to water penetration resulting from use of coupling
agents does not result from the properties of the coupling agent itself but rather from the total nature of the interface resulting from
its reaction with both glass and resin. The coupling agent anchors
the resin to the glass in such a fashion as to prevent formation of
liquid water at the interface.

REFERENCES

1. W. H. Zachariasen, J. Amer. Chem. Soc., 54, 3841 (1932).
2. R. E. Lowrie, in "Modern Composite Materials," (L. J.
   Broutman and R. H. Krock, eds.) Addison-Wesley, Reading,
   Mass., 1967, p. 280.
3. R. E. Lowrie, in "Modern Composite Materials," (L. J.
   Broutman and R. H. Krock, eds.) Addison-Wesley, Reading,
   Mass., 1967, p. 307.
4. R. K. Iler, "The Colloid Chemistry of Silica and Silicates,"
   Cornell, Ithaca, N. Y., 1955.
5. G. J. Young, J. Colloid Sci., 13, 67 (1958).
6. V. J. Deitz, private communication.
7. M. E. Schrader, unpublished results.
8. V. Ya. Davydov, A. V. Kiselev, and L. T. Zhuravlev, Trans.
   Faraday Soc., 60, (12) 2254 (1964).

9. G. A. Galkin, A. V. Kiselev, and V. I. Lygin, Trans. Faraday Soc., 60, 431 (1964).

10. C. G. Armistead and J. A. Hockey, Trans. Faraday Soc., 63, 2549 (1967).

11. J. B. Peri and A. L. Hensley, Jr., J. Phys. Chem., 72, 2926 (1968).

12. C. G. Armistead, A. J. Tyler, F. H. Hambleton, S. A. Mitchell, and J. A. Hockey, J. Phys. Chem., 73, 3947 (1969).

13. F. H. Hambleton and J. A. Hockey, Trans. Faraday Soc., 62, 1694 (1966).

14. J. A. G. Taylor and J. A. Hockey, J. Phys. Chem., 70, 2169 (1966).

15. C. G. Armistead and J. A. Hockey, Trans. Faraday Soc., 63, 2549 (1967).

16. M. L. Hair and W. Hertl, J. Phys. Chem., 73, 2372 (1969).

17. F. H. Van Cauwelaert, P. A. Jacobs, and J. B. Uytterhoeven, J. Phys. Chem., 76, 1434 (1972).

18. M. L. Hair and W. Hertl, J. Phys. Chem., 77, 1965 (1973).

19. E. G. Shafrin and W. A. Zisman, J. Am. Ceram. Soc., 50, 478 (1967).

20. T. Young, Phil. Trans. Roy. Soc., 95, 65 (1805).

21. A. Dupré, "Theorie Mechanique de la Chaleur," Gauthier-Villars, Paris, 1869, p. 369.

22. C. G. Sumner, "Symposium on Detergency," Chemical Publishing Co., New York, 1937, p. 15.

23. R. Shuttleworth and G. L. Bailey, Discussions Faraday Soc., 3, 16 (1948).

24. R. E. Johnson, J. Phys. Chem., 63, 1655 (1959).

25. D. H. Bangham, Trans. Faraday Soc., 33, 805 (1937).

26. D. H. Bangham and R. I. Razouk, Trans. Faraday Soc., 33, 1459 (1937).

27. R. N. Wenzel, Ind. Eng. Chem., 28, 988 (1936).

28. W. A. Zisman, Advances in Chemistry Series, No. 43, p. 1, American Chemical Society, Washington, D.C., 1964.

29. W. D. Harkins, "The Physical Chemistry of Surface Films," Reinhold, New York, 1952.

30. H. W. Fox and W. A. Zisman, J. Colloid Sci., 5, 514 (1950).

31. M. J. Rand, J. Electrochem. Soc., 109, 402 (1962).

32. C. H. Amberg and R. McIntosh, Can. J. Chem., 30, 1012 (1952).

33. M. E. Schrader, J. Colloid Interface Sci., 27, 743 (1968).

34. R. W. Roberts and T. A. Vanderslice, "Ultrahigh Vacuum and Its Applications," Prentice-Hall, Englewood Cliffs, N. J., 1965.

35. V. Ponec, in "Treatise on Adhesion and Adhesives," Vol. 2, (R. L. Patrick, ed.) Dekker, New York, 1969, Chapter 11.

36. S. P. Wolsky, J. Appl. Phys., 29, 1132 (1958).

37. A. J. Rosenberg, P. H. Robinson, and H. C. Gatos, J. Appl. Phys., 29, 771 (1958).

38.   L. H. Germer and C. D. Hartman, J. Appl. Phys., 31, 2085
      (1960).
39.   H. E. Farnsworth, R. E. Schlier, T. H. George, and R. M.
      Burger, J. Appl. Phys., 26, 252 (1955).
40.   H. E. Farnsworth, R. E. Schlier, F. H. George, and R. M.
      Burger, J. Appl. Phys., 29, 1150 (1958).
41.   H. E. Farnsworth, "The Surface Chemistry of Metals and Semi-
      conductors," (H. C. Gatos, ed.) Wiley, N. Y., 1960, p. 21.
42.   H. E. Farnsworth, Ann. N. Y. Acad. Sci., 101, 658 (1963).
43.   L. Holland, "Vacuum Deposition of Thin Films," Chapman and
      Hall, London, 1963.
44.   D. Haneman, Phys. Chem. Solids, 11, 205 (1959).
45.   J. J. Lander, G. W. Gobeli, and J. Morrison, J. Appl. Phys.,
      34, 2298 (1963).
46.   M. E. Schrader, J. Phys. Chem., 74, 2313 (1970).
47.   M. E. Schrader, J. Phys. Chem., 78, 87 (1974).
48.   N. Fried, private communication.
49.   W. I. Patnode, U.S. Patent 2,306,222.
50.   E. G. Rochow, "An Introduction to the Chemistry of the Silicones,"
      Wiley, New York, 2nd ed., 1951.
51.   P. W. Erickson and R. Middlebrook, NOL Technical Report
      62-63, March, 1962.
52.   R. K. Witt, in "Historical Background of the Interface-Studies
      and Theories" by P. W. Erickson, J. Adhesion, 2, 134 (1970).
53.   J. Bjorksten and L. L. Yaeger, Mod. Plast., 29, 124 (1952).
54.   J. Bjorksten, L. L. Yaeger, and J. E. Henning, Ind. Eng.
      Chem., 46, 1954 (1962).
55.   H. A. Clark, Mod. Plast., 30, 142 (1952).
56.   S. Sterman and H. B. Bradley, SPE Trans., 1, 224 (1961).
57.   M. H. Jellinek and N. D. Hanson, Mod. Plast., 35, 178 (1957).
58.   E. L. Lotz, Amer. Dyestuff Reporter, 48, 50 (1959).
59.   R. Steinman, Mod. Plast., 29, 116 (1951).
60.   N. M. Trivisonno, L. H. Lee, and S. M. Skinner, Ind. Eng.
      Chem., 50, 1912 (1958).
61.   P. W. Erickson and I. Silver, U.S. Patent 2,720,470.
62.   S. S. Oleesky and J. G. Mohr, "Handbook of Reinforced Plastics,"
      Reinhold, New York, 1964, pp. 124, 136.
63.   W. J. Eakins, Report No. 18, U.S. Plastics Technical Evalua-
      tion Center, Dover, N. J., 1964.
64.   H. Lee and K. Neville, "Handbook of Epoxy Resins," McGraw-
      Hill, New York, 1967.
65.   J. V. Duffy, Naval Ordnance Lab. Tech. Rept. 65-137, 1965.
66.   J. V. Duffy and P. W. Erickson, Naval Ordnance Lab. Tech.
      Rept. 64-204, 1965.
67.   E. P. Plueddemann, J. Adhesion, 2, 184 (1970).
68.   Y. L. Fan and R. G. Shaw, Proc. 25th Conf. SPI Reinf. Plast.
      Div., Sec. 16A, 1970.

69. P. W. Erickson, J. Adhesion, 2, 131 (1970).
70. R. Steinman, U. S. Patent 2, 688, 006.
71. D. E. Bradley, Brit. J. Appl. Phys., 10, 198 (1959).
72. M. E. Schrader, I. Lerner, and F. J. D'Oria, Mod. Plast., 45, 195 (1967).
73. M. E. Schrader, J. Adhesion, 2, 202 (1970).
74. M. E. Schrader and A. Block, J. Polymer Sci., Part C, 34 281 (1971).
75. D. I. James, R. H. Norman, and M. H. Stone, Plast. Polymer, February, 1968, p. 21.
76. S. Sterman and J. G. Marsden, Proc. 21st Conf. SPI Reinf. Plast. Div., Sec. 3A, 1966.
77. O. K. Johannson, F. O. Stark, G. E. Vogel, and R. M. Fleischman, J. Compos. Mater., 1, 278 (1967).
78. M. E. Schrader, in "Interface Phenomena in Polymer Matrix Composite Materials," (E. P. Plueddemann, ed.) Academic, New York, 1974.

Chapter 15

THE ROLE OF FRICTION IN THE MECHANICAL
BEHAVIOR OF FABRICS

Percy Grosberg

Department of Textile Industries
The University of Leeds
Leeds, England

I. INTRODUCTION

The energy loss which occurs during cyclic deformation of fabrics is due to two separate causes. The first is the well-known non-Hookean behavior of the fibers themselves, while the second is in some way related to the structure of the fabric. The basic physical mechanisms for these two causes of hysteresis are rather different; the first is the result of plasticity and creep effects in the fiber, while the second is due to frictional restraint to interfiber and interyarn movements in the fabric during deformation. The effect of both mechanisms, however, is often very similar, so that fiber hysteresis may be considered as due to "internal" friction. For our present purposes, however, we will consider only the effect of friction in the more conventional sense,

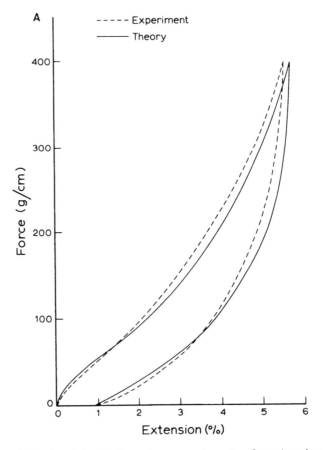

FIG. 1.   (a) Hysteresis curve for simple extension.

i. e., the frictional restraint set up when relative movements occur between yarns or fibers.

The energy loss during small deformations is due almost completely to frictional effects, since under small deformations the fibers are approximately Hookean, while the full frictional effect is usually operative at extremely small deformations. Consequently most fabrics show a three-stage behavior during deformation: (1) an initial, nonlinear stage where friction largely determines the force required to produce a fixed deformation, (2) a linear region where the elastic deformation of the fibers determines the extra force required, and (3) a nonlinear region where as a result of large deformations plasticity and creep effects in the fiber become increasingly noticeable. Figure 1 shows typical cyclic loading curves for (a) the load extension, (b) the shear, and (c) the bending of the same woven fabric for relatively small deformations. The two-stage behavior can

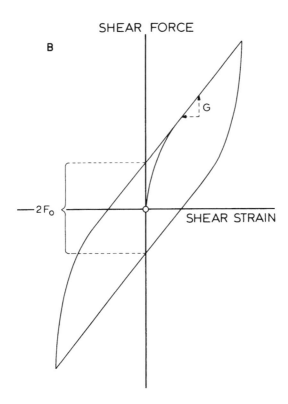

FIG. 1.   (b) Hysteresis curve for bending.

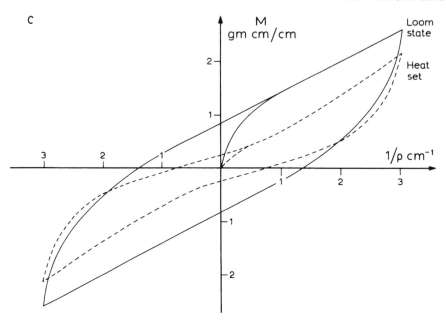

FIG. 1. (c) Hysteresis curve for shear of a woven fabric.

be seen clearly. However, for load extension behavior a second elastic region is noted because of a change in the basic mode of fiber deformation with increased fabric deformation.

This initial nonlinear deformation has been studied extensively by a number of workers [1, 2, 3, 4] who have shown the importance of this phenomenon in both laboratory tests of fabric mechanical properties and objective tests of such important fabric properties as handle, drape, and crease resistance. In this chapter it is proposed to consider (1) the basic factors which affect the frictional energy loss during fabric deformation and (2) what useful information regarding fabric end-use characteristics can be obtained from a knowledge of the size of this frictional energy loss.

II. FABRIC DEFORMATION

A. The Frictional Energy Loss During Fabric Deformation

The frictional energy is the sum of the products of the relative movement between either yarns or fibers, the force acting between them and the coefficient of friction at each contact point where

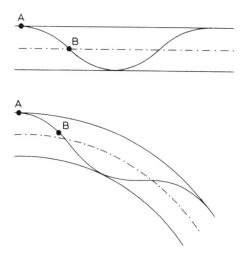

FIG. 2. Relative movement of fiber in a twisted yarn during bending.

relative movement occurs. For a given fabric deformation the relative fiber or yarn movements are clearly a function of the fabric geometry, i. e., both of the geometry of the fibers in the yarns and of the yarns within the fabric. The force acting at fiber and yarn contact points can be due to two causes: (1) internal stresses produced in the fabric during fabric and/or yarn formation and (2) internal stresses due to the fabric deformation itself. The coefficient of friction can be varied by a variety of treatments, but our quantitative knowledge of this aspect of frictional behavior in fabric deformation is relatively meager. The three basic factors will now be considered in more detail.

B. Relative Fiber and Yarn Movements During Fabric Deformation

The size of the relative yarn and/or fiber movements during load extension, shear, and bending of woven fabrics have been considered in a series of publications [5, 6, 7, 8]. It has been shown that during bending the most important relative movement is that due to the difference in displacement of fibers lying in different radial positions. As a result of yarn twist the fibers take up helical paths. When such helical paths are bent there is a movement of the fibers from that part of the helix which lies in the lower half of the bent curve to that which lies in the upper half, as shown in Figure 2. The maximum movement, which occurs at the neutral plane, varies with the helix radius so that fibers which are in contact, but occupy different radii, rub on each

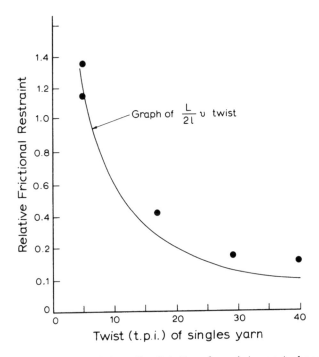

FIG. 3.   Effect of twist on the frictional resistance to bending.

other during yarn bending. It has been shown [5] that the relative movement depends on the yarn twist; the larger the twist, the smaller the frictional energy loss—other factors remaining the same. Figure 3 shows that these predictions agree closely with actual measured values for woven fabrics. Similar movements occur during the bending of knitted fabrics, but no quantitative predictions of the size of these movements has as yet been made.

The shear of woven fabrics results largely in the alteration in the angle between warp and weft yarn, so that a rotation occurs at each intersection. The relative movement between the yarns clearly depends on the area of contact between the yarns, so that once more the relative movement is determined by the fabric geometry. The size of this contact region is difficult to determine, but Figure 4 shows that extimates of the frictional hysteresis based on some probable estimates of this contact area give reasonable agreement with experimental values. It has been shown that during shearing the yarns are also bent to some extent so that interfiber movements occur but the effect of these on the total frictional energy loss is unimportant. The shearing of knitted fabrics also results in relative yarn movements,

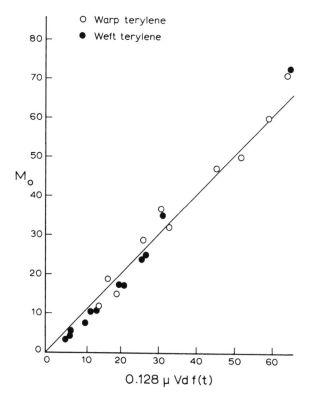

FIG. 4. Experimental and predicted value of frictional resistance to bending.

but the movement is no longer a swivelling or rotation of one yarn on another but a relative sliding of one yarn relative to another at the point there intersecting loops are in contact with each other.

During load extension the yarns at each intersection retain their relative position when the direction of loading is in either the warp or weft direction. Under these conditions, however, interfiber movements occur due to the change in yarn shape between intersections. Figure 5 shows the shape change that occurs between intersections. As can be seen, this space change implies a movement of fibers on the upper part of the curve to the left, while those on the lower part of the curve move to the right. Clearly, relative fiber movements occur which depend on the movement of fibers lying at different distances from the neutral plane. Notice that this relative movement is not affected by yarn twist, unlike that discussed above for yarn bending. This relative movement depends on the yarn geometry rather than on

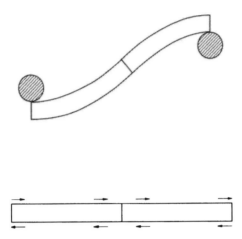

FIG. 5. Relative fiber movement during decrimping of a woven fabric during extension.

the fiber geometry. The size of this relative movement has been esti-mated [6]. Figure 1a, which has been taken from this reference, shows that by this means the hysteresis curve for small cyclic ex-tensions of woven fabrics can be accurately predicted. Similar de-formations occur in the extension of knitted fabrics.

If fabrics are deformed in some other way, for example by ex-tension in the bias direction, both interfiber and interyarn movements occur. In general, however, for any woven fabric the only movements of any importance are (1) fiber movements along the twist helix, (2) yarn rotations, and (3) fiber movements away from and toward yarn inflexion points. For knitted fabrics other relative movements are known to occur, but no exact analyses are available as yet.

C. Force Acting at Contact Points

As has previously been mentioned, the pressures between yarns and fibers arise either during manufacture, and therefore exist in the fabric before any deformation occurs, or are the result of the deforma-tion itself. It would be expected that due to both the creep behavior of most fibers and the fact that "setting" and "relaxation" treatments are applied to all fabrics during processing, the stresses which exist in the fabric before deformation would be negligible. One of the major findings of the work that has been done on frictional energy losses during fabric deformation has indicated that this is not in fact the case. The forces occurring for example at yarn intersections in a woven fabric are very small, about two orders smaller, in comparison with the forces required to make the woven fabric in the first place; but they are large in comparison with the small forces needed to bend or shear each unit cell of the fabric.

It is simple to judge whether the force at the intersection is due mainly to initial internal stresses or to stresses caused by the deformation from the shape of the hysteresis curve; since the width of the hysteresis curve at a given fabric deformation is directly proportional to the force at the intersection. As can be seen from Figure 1, this width is independent of the deformation for shear-deformation curve but varies greatly with the deformation for the load-extension curve. Two curves are shown on the bending-curve graphs. The one for loom-state fabrics and the other for a heat-set fabric. It can be seen that for loom-state fabrics the variation in width of the hysteresis curve is negligible, but it is quite noticeable for a heat-set fabric. These observations indicate that during the shear of all fabrics and the bending of loom-state fabrics, the force at fiber and yarn contacts remains constant during deformation. During warp or weft-wise extension and the bending of heat-set fabrics, the force at contacts increases as the deformation increases. It is interesting to note that theoretical analyses of all these deformations [9] indicates the same trend in the variation in force at intersections as that shown by the experimental findings. It is shown later that that part of the hysteresis which is due to the initial stress condition has more effect on the end-use characteristics of fabrics than that part which is due to the deformation-induced internal stress.

A countering force can be induced between intersections by applying another set of forces on the fabric as a whole before the forces required to produce the deformation being studied are applied. Thus it is commonplace to apply a simple tension to a fabric before shearing it to prevent buckling. The effect of such initial tensions have been studied [10]. The effect of initial, biaxial stresses prior to shearing have also been studied in some detail [11].

D. Coefficient of Friction

It has been shown experimentally [8, 12] that by the addition of lubricants or "softening" agents the coefficient of friction can be modified and the frictional hysteresis reduced accordingly. The size of the modifications in the frictional resistance to mechanical deformation which can be produced in this way is relatively small, especially when compared to the effect produced by stress-relaxation techniques such as heat- or chemical-setting treatments. Thus, a 50% reduction in frictional resistance to deformation by the addition of softening agents is as large a change as has been noted in practice. A 100% reduction from loom state to well-finished fabric, on the other hand, is commonplace.

To summarize our knowledge regarding the b asic factors which determine the frictional resistance to mechanical deformation, it has been shown that there are three factors. The first, which is determined by the yarn and fiber geometry, is the ratio of the relative yarn or fiber movement to the fabric deformation. The second, which is largely determined by the state of unreleased manufacturing stress in the fabric, is the force between yarns and fibers at intersections. The third factor, which is dependent on the fiber type and surface characteristics as well as the presence of "softening" agents or other lubricants, is the coefficient of friction between yarns and fibers.

III. EFFECT OF FRICTIONAL RESTRAINT ON THE END-USE CHARACTERISTICS OF FABRICS

Before considering this topic it is necessary to have a quantitative measure of the frictional restraint. Although there is no standard method for defining the frictional restraint, the majority of workers in the field have expressed the frictional restraint in bending and shear by the half-width of the hysteresis curve at zero deformation, viz. the half the width $2F_o$ in Figure 1(c). This value has been termed the coercive couple, M_o, in fabric bending and has been denoted by F_o in fabric shear. Since fabric compression is not possible it is difficult to measure a similar value in fabric load extension.

The elastic resistance to bending and shear has been quantified by the slope of the straight line portions in Figure 1(c), denoted by G in bending and B in shear.

The connection between these values and the basic physical factors affecting the frictional resistance to shear and bending can be summarized by the equations

$$M_o = 0.128 \, \mu_1 VdL/2l$$
$$F_o = 3 \, \mu_2 Vd_c/16(p_2 l_1 + p_1 l_2)$$

where

μ_1 = coefficient of friction between fibers

μ_2 = coefficient of friction between yarns

V = total force acting between yarns at an intersection

d = yarn diameter

d_c = diameter of contact area between yarns

l = length of yarn in unit repeat of woven structure (i.e., the modular length) p = distance between intersections

L = distance between twist reversals (i. e. , length per unit twist)

Note: subscripts 1 and 2 apply to the warp and weft directions, respectively.

The only parameter in these equations which cannot normally be measured is V, the force acting at yarn intersections. In fact, one of the main reasons for measuring fabric shear and bending is to monitor the changes in V that occur during finishing procedures. V, as has been pointed out, arises from either mechanical loads applied to the fabric, or from the state of stress in the fabric. This state is mainly governed by finishing routines. The production of certain desirable fabric "aesthetic" characteristics is in some way related to the frictional resistance to bending and shear. This is largely brought about by the skill of the finisher in altering the internal stress and to a smaller extent the interfiber and interyarn friction. The relationships between these fabric end-use characteristics and the fabric mechanical behavior is complex, and not yet completely explored, but some known facts will be mentioned.

A.   Handle

It has been shown in one series of trials (using finishers as the assessors of the handle of a series of fabrics) that there was a strong correlation between handle and the coercive or frictional couple Mo [13]. In a more widespread but less scientifically controlled test where fabric buyers were used as the assessors of fabric handle, it would appear that the correlation was in fact with the ratio of Mo to B rather than Mo alone. In this unpublished survey it would seem that fabrics with an Mo:B value less than 0.3 cm^{-1} were considered to have a soft handle, between 0.3 and 0.6 cm^{-1} a crisp handle and above 0.6 cm^{-1} a harsh handle. Some further work has suggested a more complex relation between handle and Mo and B [14].

Clearly the frictional restraint to bending has an effect on bending, but large scale controlled tests are needed in this field to establish the exact correlation.

B.   Drape

The shape and form of the drape formed in two dimensional buckling, e. g. , the drape of a curtain, has been shown to be a function of the ratio Mo:B [15] only. Three-dimensional drape, as produced by the drapeometer is more complex, but it is affected by the initial resistance to bending and shear [16] of the fabric. These resistances, as has already been pointed out, are affected by the frictional restraints in the fabric to such deformations. But while some of the laboratory techniques for measuring drape have been extensively researched, so that we do know that friction affects these methods of measurement, the question as to what aspects of drape are considered to be desirable in different end-uses remains relatively uncharted territory.

C. Wrinkle and Crease Resistance

The relative effect of frictional restraint to bending compared with fiber plasticity and creep on the wrinkle and crease resistance of fabrics depends on the relative severity of the creasing or wrinkling action. In severe creasing the frictional resistance plays only a minor role in determining the crease resistance. Many authors therefore feel that the contribution made to crease resistance in use by a reduction in frictional resistance to bending is of small importance. However, it has been shown [14] that the frictional resistance does contribute to the values of crease resistance as obtained by some of the standard methods of measurement.

IV. CONCLUSIONS

The bending shear and load extension of fabrics involves relative movements between fibers and/or yarns which are in contact. As a result there is always a frictional component to the resistance of a fabric to deformation. This frictional resistance can be modified by standard finishing procedures, and it is clear that these changes affect what, for want of a better word, can best be described as the aesthetic properties of the fabric. While it is possible to monitor finishing procedures by measuring the frictional resistance to bending or shear, the extra knowledge of the precise properties of the fabric, which provide a certain required handle or drape, is not available as yet. Therefore it is not possible to "engineer" fabrics in this particular sense. The much more precise knowledge which we now possess on the nature of the resistance to deformation of fabrics indicates that such knowledge will soon be available.

REFERENCES

1.   T. Eeg - Olofsson, J. Text. Inst. , 50, T112 (1959).
2.   J. D. Owen, J. Text. Inst. , 59, 313 (1968).
3.   J. Lindberg, "The Setting of Fibres and Fabrics, " Merrow, 1971, p. 166.

4. Vivien H. Dawes and J. D. Owen, J. Text. Inst., 62, 181 (1971).
5. G. M. Abbott, P. Grosberg, and G. A. V. Leaf, Text. Res. J., 41, 345 (1971).
6. J. W. S. Hearle, P. Grosberg, and S. Backer, "Structural Mechanics of Fibres, Yarns and Fabrics," Vol. 2 (in press).
7. P. Grosberg and B. J. Park, Text. Res. J., 36, 420 (1966).
8. P. Popper, Sc. D. Thesis, Massachusetts Institute of Technology, 1966.
9. G. M. Abbott, P. Grosberg, and G. A. V. Leaf, J. Text. Inst., 64, 346 (1973).
10. P. Grosberg, G. A. V. Leaf, and B. J. Park, Text. Res. J., 38, 1087 (1968).
11. S. Kawabata, N. Masako, and H. Kawai, J. Text. Inst., 64, 62 (1973).
12. P. Grosberg and J. W. Rhee, Applied Polymer Symposium 18, 1303 (1971).
13. P. Grosberg, "Studies in Modern Fabrics," The Textile Institute, 1970, p. 112.
14. Vivian H. Dawes and J. D. Owen, J. Text. Inst., 62, 233 (1971).
15. P. Grosberg and N. M. Swani, Text. Res. J., 36, 332 (1966).
16. G. E. Cusick, J. Text. Inst., 55, T596 (1965).

Chapter 16

SURFACE PROPERTIES IN RELATION TO
THE BONDING OF NONWOVENS

G. G. Allan, J. E. Laine, and A. N. Neogi

Department of Chemical Engineering
College of Forest Resources
University of Washington
Seattle, Washington

I. INTRODUCTION

Nonwovens usually consist of at least two components; the fiber and the binder. Obviously, the characteristics of this combination must be primarily derived from the physical and chemical properties of the individual components and their mutual interactions, which are also a

function of their arrangement in space [1]. Although this large number of variables implies that any particular nonwoven is likely to be
relatively unique, such special structures can be related to a fundamental parent which is susceptible to precise description [2]. This
progenitor consists of a random web of fibers of equal length, bonded
by discrete spots of binder distributed throughout the web in a random
fashion. While such a model nonwoven may appear at first sight to be
of limited and hypothetical interest, in fact many commercial products show only a slight divergence from this idealized structure [3].
As a specific example, conventional paper consists of an essentially
random web of fibers, albeit with some fiber alignment in the machine
direction, each of which is bonded to another at discrete intersections
by hydrogen bonds [4]. Similarly, air-deposited webs that are bonded by saturation-impregnation techniques [5] can be regarded as a
randomly spot-bonded composite in which the binder spots are infin-
itely close. Likewise, pattern-bonded nonwovens [6-10] which are
already important articles of commerce, can be viewed as distant
cousins of paper where the distribution of binder sites is no longer
random but must be defined by some other mathematical function
[11]. Our understanding of the effective role of bonding on many
types of nonwovens can therefore be augmented by analytical consideration of the basic model structure.

Thus, since the function of the binder in a randomly bonded random fiber network is to ensure the transference of load from one fiber to the next, the tensile strength, T, of such a composite is given
by

$$T = fZ \tag{1}$$

where Z is the zero-span tensile strength of the assemblage [12] and
f is the fraction of fibers in the web which are interconnected by binder to other fiber segments at least twice and are hence capable of
sharing the load. Since the zero-span tensile value is obtained by
rupture of the nonwoven under conditions where only fibers break,
Z is a reflection of the physical properties of the fiber content of the
web. In contrast, f is a function of the content of binder and its distribution within the fiber network. If the binder amount and arrangement are such that on the average each fiber contacts more than four
bond sites, then almost all the fibers share the load and $f \to 1$ so that

$$T = Z \tag{2}$$

Equation (2) only applies if all the fibers along the failure line break simultaneously. This however is a rare occurrence. In practice, as
a result of poor binder-fiber adhesion or a lack of adequate contact
area of bonding, many fibers slip and pull out of the enveloping binder
under the application of tensile stress [13-15]. This fiber slippage
causes a catastrophic failure at a later stage and, as a consequence,

the tensile strength of the nonwoven is always less than predicted. Since the fibers pulled out carry no load, Eq. (1) can be rewritten so that

$$T = fZ(1 - f_p) \tag{3}$$

where f_p is the fraction of fibers pulled out across the future line of failure. For a randomly bonded random fiber web, it can be shown that [16] f_p is a function of the ratio of fiber tenacity (σ_f) to bond strength $(\sigma_b \ell_b)$ and is quantitatively given by

$$f_p = \frac{14\sigma_f}{9\ell_b \sigma_b} \tag{4}$$

where σ_b is the bond strength per unit length of fiber bonded and ℓ_b is the average bonded length per fiber. Thus an expression for the tensile strength can be obtained in terms of these fundamental parameters by combination and rearrangement of Eqs. (3) and (4) so that

$$T = fZ \left[1 - \left(\frac{14\sigma_f}{9\sigma_b \ell_b} \right) \right] \tag{5}$$

Since $f \to 1$ when bonding is extensive, Eq. (5) can be simplified and transformed into the reciprocal relationship

$$\frac{1}{T} = \frac{1}{Z} + \frac{1}{B} \tag{6}$$

where B is an index of bonding. Now, since the tensile strength of a random fiber assemblage has also been related [16] to the areal fiber density of the web by the expression

$$T = \frac{\sigma_f n_f}{3} \tag{7}$$

where, in this case, n_f is the number of fiber centers per unit area; it therefore follows that

$$B = \frac{3\sigma_b \ell_b n_f}{14} \tag{8}$$

Examination of Eq. (6) illuminates the electrical analogy that the resistance of the nonwoven to tensile rupture is the sum of two resistances in parallel; the resistance of fibers to failure under tension [2] and the resistance of bonds to shear failure.

Furthermore, Eq. (6) also defines the limits achievable by manipulation of the properties of the constituents of the nonwoven. That is, improvement of the bond strength can only increase the tensile strength of the structure until the limiting value of Z is reached. Comparably, augmentation of fiber strength has an effect limited in turn by the value of the factor B. This latter restriction is particularly valid in practice where the substitution of polyester (6g/den) for rayon (2.6g/den) fibers does not lead to correspondingly stronger nonwovens because the strength of the assemblage is bond-strength limited. These theoretical and practical considerations clearly emphasize the importance of bond strength in a composite and moreover indicate that the key to the improvement of bond strength lies in the product, $\sigma_b \ell_b$. Now the bond shear strength per unit bonded length, σ_b, is a composite term which can be separated into two components; the shear strength of adhesion, σ_a, and the perimeter of adhesion, P_a. It therefore follows that

$$\sigma_b \ell_b = \sigma_a P_a \ell_b \tag{9}$$

The shear strength of adhesion, σ_a, is of course a fundamental property for any particular binder-fiber duo, determined by their mutual chemical compatibility. In contrast, the bonded length, ℓ_b, is a parameter principally influenced by the physical mode of binder distribution.

Similarly, the perimeter of adhesion, P_a, is also a physical property and is directly related to geometry of the fiber cross-section at the fiber-binder confluence. That is, for fibers of the same denier a triangular cross-section offers substantially more area for adhesion to an enveloping binder than, for example, a fiber with a circular cross-section. Thus, the bonding strength and efficiency for a given amount of binder in a nonwoven is clearly dependent on the surface properties of the fiber in terms of (a) topology and (b) chemical interaction with the binder.

II. TOPOLOGICAL CONSIDERATIONS IN BONDING

A. Topology of Fiber Surfaces in Relation to Binder-Fiber Interactions

The cross-sectional shape of a fiber is subject to infinite variation, and the anticipated effects of some of the standard geometries in non-woven structures has been recently discussed [17, 18] in terms of

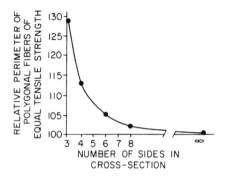

FIG. 1. Relative perimeter of fibers of equal strength and regular polygonal cross-section.

overall flexibility and bonding. Since the bond strength is equal to the force necessary to pull out a fiber from the binder matrix, it should be clear that the bond strength is directly proportional to the perimeter of the fiber, P_f. Indeed, this has been experimentally verified by Hearle and Newton [19]. Therefore, Eq. (9) can be recast in the form

$$\sigma_b \ell_b = K \sigma_a P_f \ell_b \tag{10}$$

where K is a constant of proportionality. Consequently the importance of perimeter dimensions should not be underemphasized in evaluations of binders because, as noted earlier, nonwoven structures frequently fail by fiber slippage through the binder matrix. It therefore follows that, other things being equal, the strength of a nonwoven structure based on circular fibers should be less than, for example, a similar nonwoven structure composed of fibers of equal strength with a triangular cross-section which has a perimeter 28% greater in length. This is, of course, a specific example; in general terms the perimeter of a fiber of regular polygonal cross-section is given by

$$P_f = 2 \left(nA \tan \frac{\pi}{n} \right)^{1/2} \tag{11}$$

where A is the cross-sectional area and n designates the number of sides. The effect of changing the polygonal shape of a fiber can be seen by plotting the relative perimeter of a series of regular polygons of constant area against the number of sides. As is evident in Figure 1, the curve begins to rise slowly as the number of sides decreases from infinity, for the circular cross-section, until the octagonal geometry is reached. Obviously the bonding benefits of switching from fibers with a circular cross-section to fibers with an octagonal cross-section

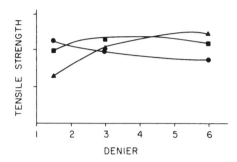

FIG. 2.   Variation of tensile strength (arbitrary units) in nylon 6, 6 composites with fiber denier and length; 6. 4 mm (●), 9. 5 mm (■) and 12. 7 mm (▲).

would be insignificant.   In contrast, after the octagonal fiber, the relative perimeter starts to increase rather rapidly until the maximum associated with the triangular cross-section is reached.   Of course, the perimeter of a fiber cross-section can be changed without changing its shape.   That is, the fineness can simply be increased or decreased. The foregoing arguments would of course still be applicable and the comparison by Hentschel [20] of nylon webs of different deniers are in good accord with this view.   The data in Figure 2 shows that, for webs of equal weight, the finer fibers yield the strongest nonwoven at a given binder level for a fiber length of 6.4 mm.   This can be attributed to the total increase in fiber surface interacting with the binder.   Longer fibers, on the other hand, apparently do not exhibit this tendency, although the area of interacting surface certainly increases as the fiber denier decreases.   This anomalous behavior can probably be ascribed to fiber entanglement, which becomes more severe the longer the fiber and thereby reduces web uniformity.

A curve somewhat analogous to Figure 1 is shown in Figure 3, where the relative perimeters of a series of regular polygonal fibers of identical flexibility and consequently differing areas of cross-section are plotted against the number of sides.   The moments of inertia of the fibers, I, with the various cross-sections were calculated from the standard relationship [18]

$$I = \iint y^2 \, dy \, dx \tag{12}$$

Figure 3 illustrates that in this situation, where the stiffness of all the fibers is axiomatically identical, the adhesive perimeters of the lower polygons are surprisingly and significantly greater than the limiting circular case.   It is especially noteworthy that the fiber with the triangular cross-section has still almost 23% additional surface per unit length relative to the circular fiber of the same stiffness.

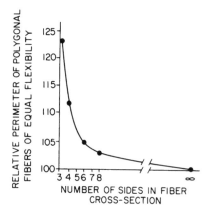

FIG. 3. Relative perimeter of fibers of equal flexibility and regular polygonal cross-section.

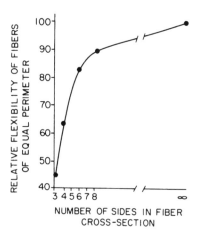

FIG. 4. Relative flexibility of fibers of equal perimeter and regular polygonal cross-section.

An intriguing consequence of this geometrical phenomenon follows from a scrutiny of the comparative flexibilities of regular polygonal fibers of constant perimeter. From the data collected in Figure 4 it is apparent that if these fibers are bonded in such a way that an equal force is required to shear the adhesive bond in each case, then the triangular fiber complex will be the most flexible structure and the circular fiber assemblage the most rigid. However, this initially somewhat startling conclusion can be more readily accepted when it is appreciated that the cross-sectional area of the triangular fiber is only about 60% of that of its circular counterpart.

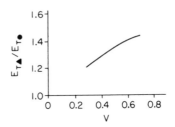

FIG. 5. Variation of transverse stiffness of epoxy composites from glass fibers with triangular $(E_{T_{\blacktriangle}})$ and circular $(E_{T_{\bullet}})$ cross-sections. V is matrix volume percent.

These theoretical arguments serve to emphasize the importance of fiber cross-sectional shape in the bonding of nonwoven structures. Experimental support for this conclusion is provided by the data of Broutman and Krock [21] depicted in Figure 5. A comparison of otherwise identical triangular and circular fibers in an epoxy matrix demonstrates dramatic differences in physical properties.

Although the production of fibers of noncircular cross-section imposes certain difficulties, the technology to accomplish this has been established and examples of such fibers shown in Figure 6 are already available commercially [22, 23].

So far only fibers with geometrically regular polygonal cross-sections have been considered. Obviously, much more impressive increases in the perimeter dimensions can be obtained by the creation of asymmetric fiber cross-sections. The relative perimeters of a number of such fibers are summarized in Table 1, where it can be seen that fibers with cross-sections in the shape of an I-beam have a potential bonding area 90% greater than fibers of equal strength with circular cross-sections. Certainly it can be anticipated that fibers with the more exotic and highly bondable cross-sectional shapes depicted in Figure 7 will become more and more important in nonwovens as their availability expands [24].

A fiber cross-section may also vary in area along the fiber axis, as shown in Figure 8, offering another possible bonding dimension in the future for nonwovens whose strengths are bond limited. Such fibers, which have recently been described in the patent literature [25], should resist the tendency to pull out of an enveloping binder much as a screw anchors itself in wood. This has been demonstrated by Allan and McConnell [26], using as fiber models a series of aluminum rods, sinusoidally machined and embedded in a wax matrix. The force required to pull out the fiber model from the matrix model increases with the amplitude and frequency of the sinusoidal surface, as illustrated by the data in Figure 9.

FIG. 6.   Tracings of photomicrographs of commercially available fibers with star-shaped and quasi-triangular cross-sections.

Bond strength in a nonwoven can also be increased by the use of hollow fibers [18]. These fibers again have larger perimeters than solid fibers of equal strength and this provides for improved binder-fiber interaction. The gain in surface area for fibers of increasing fistularity is graphically expressed in Figure 10. However, it must be remembered that the stiffness of the fiber is also climbing at the same time; and, much more rapidly. This of course could be very undesirable in many applications, although when the total fiber cross-section has less than 50% solid area, spontaneous or induced fiber

TABLE 1

Relative Perimeters of Fibers of Equal Cross-Sectional Area

Description of cross-sectional shape	Relative perimeter
Circle	100. 0
Regular octagon	102. 8
Regular hexagon	105. 1
Square	113. 0
Equilateral triangle	128. 2
Solid trilobal	142. 9
Hollow trilobal	144. 5
Ellipse	109. 3
Hollow circle	115. 5
Hollow square	130. 5
Hollow triangle	148. 7
Hollow ellipse	126. 2
Regular star	148. 7
I-Beam (thick)	163. 1
I-Beam (thin)	196. 5
Horizontal rectangle	130. 5
Vertical rectangle	130. 5
Flattened triangle	140. 4
Dumbbell	162. 6
Peanut	130. 0
Symmetrical bilobal	141. 6
Unsymmetrical bilobal	111. 9

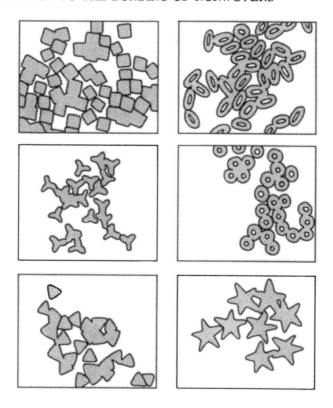

FIG. 7. Tracings of photomicrographs of fibers with profiled cross-sections.

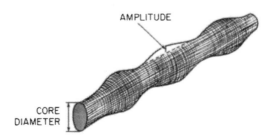

FIG. 8. Idealized representation of a longitudinally profiled fiber.

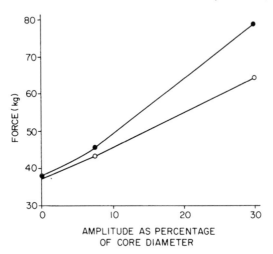

FIG. 9. The force required to pull out from a wax matrix sinu-soidally machined aluminum rods with frequencies of 39.4 m^{-1} (o) and 78.7 m^{-1} (●).

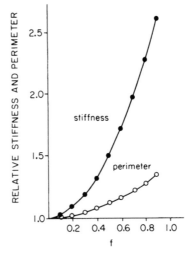

FIG. 10. Variation of stiffness and perimeter of hollow fibers rel-ative to a solid fiber of equal tensile strength.

collapse in the final nonwoven is not difficult [27]. Thereafter the fi-ber has a flat, ribbon-like cross-section which has at least the same, or even a greater bonding potential than the hollow fiber from which it originated. The increased flexibility, conformability, as well as im-proved packing, as shown in Figure 11 can also well lead to more high-ly bonded structures, particularly in the paper field [28] (Fig. 12).

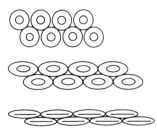

FIG. 11. Idealized representation of variation of fiber flexibility, covering power and packing ability with the fiber collapse.

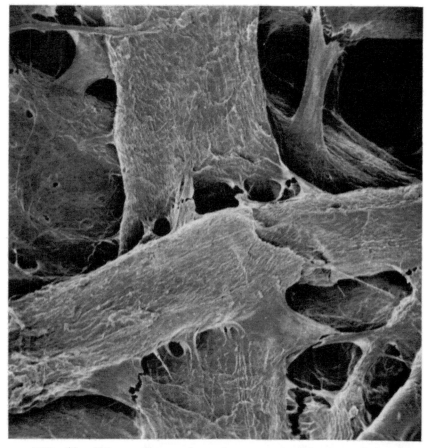

FIG. 12. Fiber collapse shown in a scanning electron photomicrograph of a chitosan-bonded handsheet prepared using an unbleached kraft pulp made from a mixture of (85:15) Western red cedar (Thuja plicata) and Western hemlock (Thuja heterophylla).

This collapsibility of hollow fibers is dependent upon the fistularity of the fiber and the geometric configurations of the inner and outer walls. For circular fibers the critical pressure which will cause collapse, P_{cr}, is given by

$$P_{cr} = 2E\left(\frac{h}{d}\right)^3 \tag{13}$$

where E is the modulus of elasticity, h is the wall thickness of the fiber and $d = (d_c + d_s)/2$ is the average of the core and surface diameters. The corresponding critical pressure relationships for hollow square (P_s) and triangular (P_t) fibers are

$$P_s = \frac{2\pi^2 Eh^3}{3 L_d \ell_d^2} \tag{14}$$

$$P_t = \frac{4\pi^2 Eh^3}{3\left(3 L_d \ell_d^2\right)^{1/2}} \tag{15}$$

where L_d and ℓ_d are the outer and inner dimensions of the fiber respectively. The combined effects of hollowness and shape on fiber adhesivity are shown in Table 2 for fibers of equal strength. Again the triangular fiber offers the most bonding potential and, although it is also the least flexible, it is easily collapsible at higher fistularities. Conversely, when the same fiber shapes are compared at equal bonding potential, as in Table 3, the triangular hollow fiber is most flexible, although the difference from the square fiber is slight.

It should be noted that the gain in surface area or bonding potential is slight for fibers of 10% hollowness in comparison to the corresponding solid circular fiber. However, at higher degrees of hollowness the potential adhesivity increases substantially and at the 95% level the circular and triangular fibers have greater than 300 and 400% more bondable surface respectively.

Finally, it must be vigorously emphasized that in all the foregoing the fibers have been discussed as if pristine geometric entities. Patently this is not the case, and any consideration of the surface properties of real fibers and nonwovens must recognize the existence of a multiplicity of surface discontinuities. The effects of this molecular-sized substructure on bonding will be reviewed later in this chapter.

TABLE 2

Comparative Properties of Hollow Fibers of Equal Strength

Hollow fraction (%)	Fiber shape	Relative buckling pressure[a]	Relative adhesivity[b]	Relative stiffness[c] (%)	Bulk
10	Circular	7,626	1.05	121	3.75
	Square	72,815	1.19	128	3.75
	Triangular	15,713	1.36	151	3.75
50	Circular	276	1.41	300	6.77
	Square	1,130	1.60	313	6.77
	Triangular	251	1.82	361	6.77
90	Circular	1.0	3.16	1895	33.85
	Square	3.4	3.57	1986	33.85
	Triangular	0.7	4.07	2313	33.85

[a]Pressures relative to critical buckling pressure of 90% hollow circular fiber.
[b]Hollow fiber perimeters relative to solid circular fiber.
[c]Moments of inertia relative to solid circular fiber.

TABLE 3

Comparative Properties of Hollow Fibers of Equal Adhesivity

Hollow fraction (%)	Fiber shape	Relative strength (%)	Relative buckling pressure[a]	Relative stiffness[b] (%)	Bulk
10	Circular	90.0	9,619	78.1	1.11
	Square	7.2	71,441	0.7	11.08
	Triangular	5.5	15,783	0.6	11.06
50	Circular	50.0	274	75.0	2.00
	Square	4.0	1,122	0.5	20.07
	Triangular	3.1	251	0.4	19.91
90	Circular	10.1	1.0	19.0	10.01
	Square	0.8	3.4	0.1	97.00
	Triangular	0.6	0.7	0.1	99.56

[a]Pressures relative to critical buckling pressure of 90% hollow circular fiber.
[b]Moments of inertia relative to solid circular fiber.

B.   Topology of Binder Conformations in Relation to Binder-Fiber Interactions

Before considering the specific question of adhesion to fibers, the distribution of binder within fibers must be established. Clearly, if every fiber intersection is made to bond, then the end-products are strong structures, similar to paper, with little textile-like quality. To build in drape, some fiber slippage is required, and this implies that every fiber intersection should not be bonded. This implication in turn directs attention to the quantitative question of the magnitude of the fraction of fiber intersections which should cohere. Some insight into this problem has been attained by grinding and size-classifying binders and bonding an air-deposited random web with the fractions applied by shaking from a pepperpot. The best combination of strength and drapeability was achieved with the material passing through a 40-mesh screen and retained on a 60-mesh screen. This type of binder arrangement within a mass of fibers has been quite extensively investigated and is the subject of a number of patents [6, 7]. These cover a variety of shapes and forms, including granules and rods of polymer or straight and wavy lines printed onto the fiber mat using polymer solutions or latices. Although these patents indicate that flexibility can be attained by such techniques, precise definition of the relationships between spacing and strength and flexibility has been worked out only recently by Allan and Neogi [2]. In this approach it is assumed that the number, size and pattern of distribution of the binder spots will determine the probability of interlinkage of the fibers within the web. To optimize these variables for the preferred combination of strength and flexibility, the number of fibers contained by a binder spot, n_t, must be characterized. If n_f is the

number of fiber centroids in a unit area of a random web and r_b is the

radius of the circular binder spot distributed randomly within the web, then it can be shown that

$$n_t = n_f \pi r_b^2 (1 + Q) \tag{16}$$

where

$$(1 + Q) = \frac{2}{\pi} (a^2 + 2a)^{1/2} + \frac{2}{\pi} (a + 1)^2 \sin^{-1} \frac{1}{(a + 1)}$$

and a is the ratio of the fiber length to binder-spot diameter. Furthermore, if the value of the ratio a is large, then the term $(1 + Q)$ in Eq.

(16) can be replaced by $2(\ell + 2r_b)/\pi r_b$ or even $2\ell/\pi r_b$ if $\ell/2r_b \gg 1$ so that

$$n_t \simeq 2n_f r_b (\ell + 2r_b) \tag{17}$$

or

$$n_t \simeq 2n_f r_b \ell \tag{18}$$

where ℓ is the fiber length.

These theoretical equations are of value for the prediction of the consequences of varying binder spot size or fiber length. Thus, reference to entries 1-5 in column 6 of Table 4 shows that, if the binder-spot size is held constant, then the number of fibers adhesively embedded theoretically increases as the length of the fibers in the web is augmented when the number of fiber centers per unit area remains constant. This is another way of expressing the accepted intuitive notion that longer fibers in a heavier nonwoven contact more binder spots. However, if the accretion of the basis weight due to the use of longer fibers is avoided by simply reducing the number of fibers in the web, then entries 2, 3, 4, and 5 in column 7 of Table 4 show that the number of fibers embedded per binder spot becomes essentially constant, since $n_f = k/\ell$ where k is a constant of proportionality. On the other hand, if the fiber length is kept invariant, then it is apparent from entries 2, 6, 7 and 8 in column 8 of Table 4 that more fibers are contacted per unit area of adhesive when the binder spot is diminished in size.

Although Eq. (16) defines the relationship between the size of the binder spot and the number of fibers embedded therein, the tensile properties of such spot-bonded nonwovens remains unspecified, although the strength is known to be related to the size and number of the spots. This relationship is illustrated by the data in Figure 13, which was derived from tensile tests of a series of rayon nonwovens bonded with polyvinyl acetate spots of different sizes. At this constant binder content it is clear that, as the size of the binder spot increases, the nonwoven initially exhibits a sharp decrease in tensile strength and ultimately falls to zero at a spot size determined by the length of the fibers. This critical spot size defines the point of loss of nonwoven coherency for fibers of this particular length. Obviously, since freedom of fiber movement is synonymous with drape, this point also establishes the level of maximum drape of the coherent nonwoven. At any higher level of binder the strength may be greater, but the fabric-like qualities will certainly be less.

ALLAN, LAINE, AND NEOGI

TABLE 4

Theoretical Prediction of Number of Embedded Fibers in a Spot-Bonded Random Fiber Assemblage

Binder spot radius r(mm)	Fiber length ℓ(mm)	Ratio of fiber length to binder spot diameter $\ell/2r = a$	$(1+Q)$	Area of binder spot (mm²)	Number of fibers embedded per binder spot per unit		
					number of fibers $n_t/n_f = \pi r^2(1+Q)$	basis weight $n_t/k = \pi r^2(1+Q)/\ell$	binder area $n_t/k\pi r^2 = (1+Q)/\ell$
1	2	1	2.44	π	7.65	3.82	1.220
1	10	5	7.56	π	23.7	2.37	0.756
1	20	10	13.7	π	43.0	2.15	0.685
1	40	20	26.9	π	84.2	2.10	0.067
1	100	50	65.6	π	206.0	2.06	0.065
2	20	5	7.56	4π	95.0	4.75	0.378
4	20	2.5	4.4	16π	221.0	11.05	0.220
10	20	1	2.44	100π	765.0	38.20	0.122

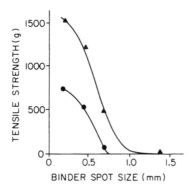

FIG. 13.   The effect of binder-spot size on the tensile strength of a rayon, nonwoven strip (15 mm) consisting of 6 mm (●) or 12 mm (▲) fibers (0. 65g) bonded with polyvinyl acetate particles (0. 2g).

This experimentally derived concept that a nonwoven, for coherency, requires a certain minimum number of bonds at each binder loading is an important one in the design of nonwovens because binder selection can thereby be rationally approached.   The concept can also be derived theoretically by consideration of any one binder spot within a randomly bonded nonwoven.   Since the maximum length of fiber which can emanate from this spot is ℓ, then the potential total number of bonding sites, N, that this particular fiber could encounter is determined by its length and the density of bonding per unit area n_b, and it therefore follows that

$$N = n_b \pi \ell^2 \qquad (19)$$

Moreover, since the binder spots are randomly distributed within the nonwoven, the probability, $p(i)$, that there are i bonds within the area reachable by the fiber, $\pi \ell^2$, is given by the Poisson distribution function:

$$p(i) = \frac{N^i e^{-N}}{i!} \qquad (20)$$

For the nonwoven to be coherent, the formation of an infinite network of fibers joined by binder spots must be created within the web.   This situation is analogous to the phenomenon of gelation in nonlinear polymerizations where the number of fibers emanating from a binder spot corresponds to the functionality of the monomer.   Consequently, if three fibers emanate from a binder spot, each fiber which reaches another binder spot can be succeeded by two more fibers.   If both of these in turn reach two more binder spots, four more fibers will be involved, and so on. However, if the probability of a fiber reaching a second binder spot is less than 1/2 there is a less than even chance that each fiber will lead to a binder spot and thus to two more fibers. That is, there is a greater than even chance that the fiber will not reach a second binder spot. Under these circumstances the network cannot continue indefinitely and a coherent structure is not possible because a preponderance of fibers must eventually build up which do not reach the next binder spot. When the probability of a fiber contacting the next binder spot is greater than 1/2 each fiber has a better than even chance of leading to two new network fibers. Two such fibers will in turn on the average lead to four new network components, and so on. Under these circumstances the network may be continued indefinitely and a coherent structure secured. Hence a probability of 1/2 that bonded fibers reach another bonding spot is the critical condition for incipient formation of a coherent fiber network. In this particular case this requirement in turn demands that there be another two bonding sites within the area reachable by the length l of the fiber used, i. e., there must be at least three spots (the original bond considered and two others) within the area πl^2 generated by the fiber pivoting in the plane of the sheet around its own end embedded in a binder spot. Now, the probability that there are less than three bonds p (<3) is the sum of the probabilities for zero, one, and two bonds, i. e.,

$$p(<3) = p(0) + p(1) + p(2) \tag{21}$$

Therefore, the probability $p(3)$ that there are at least three bonds is

$$p(3) = 1-[p(0) + p(1) + p(2)] \tag{22}$$

Using Eq. (20) to evaluate this probability it follows that

$$p(3) = 1-e^{-N}-Ne^{-N} - \frac{Ne^{-N}}{2} \tag{23}$$

Now, as shown above, for the web to be coherent $p(3)$ must be equal or greater than 1/2 so that

$$p(3) = \frac{1}{2} = 1-e^{-N}\left(1 + N +\frac{N^2}{2}\right) \tag{24}$$

from which it can be calculated that N = 2.7. Therefore, from Eq. (19), the average number of binder spots per unit area n_b can be expressed as

$$n_b = \frac{2.7}{\pi l^2} \tag{25}$$

Equation (25) reveals that the minimum number of binder spots necessary to impart coherency to a nonwoven structure is inversely proportional to the square of the length of its constituent fibers. The validity of this conclusion can be demonstrated by the construction of a series of wet-laid rayon nonwovens bonded with polyvinyl acetate beads. From the tensile properties of these nonwovens, summarized in Figure 13, the maximum size of binder spot which still allows attainment of coherency can be estimated by extrapolation. The results obtained experimentally, although small in number, do agree satisfactorily with the theoretical values. Equation (25) is therefore useful for the estimation of the minimum amount of binder necessary to secure the coherency of a nonwoven, or definition of the maximum permissible binder-particle size for nonwoven coherency at a given binder content.

III. CHEMICAL CONSIDERATIONS IN BONDING

A. Chemical Aspects of Binder-Fiber Interactions

After the fiber selection, with all its ramifications, has been debated and confirmed, attention must be turned to evaluation of the candidate binders. This is an extremely important process because the properties of the nonwoven are governed by those of the weakest link, and in all cases the binder component is an unoriented polymer and is inherently weaker than the fiber. However, if the binder is sufficiently tough and elastic to deform and accommodate itself to the forces applied and is suitably placed, failure can occur by fiber rupture. Binder characteristics leading to this condition, where a high binder resistance to shear is mandatory, are toughness and good adhesivity to the fiber. The former is usually associated with rubbery, amorphous polymers often constructed by co- or terpolymerization [5].

The difficulties of finding binders for flexible nonwovens are much greater than the problems of selecting polymers for more rigid nonwovens, and consequently attention is usually riveted to this area. Organization of a large quantity of data on the stiffness of polyethylene terephthalate nonwovens bonded by a saturation-impregnation technique with a wide range of binders gave the reasonable correlation depicted in Figure 14. Clearly then, to construct drapeable nonwovens, binders of low modulus should apparently be employed. This conclusion was later confirmed and extended by Hearle and Newton [29] in a study of the effect of binder stiffness and content on fabric modulus (Fig. 15). The level of binder has a large influence because it modifies the character of the network, curtailing the free-fiber length by diminishing the distance between bonds. In contrast, the stiffness of the binder is relatively without influence on the rigidity of the nonwoven until the binder modulus exceeds about 200g/tex. Thereafter the sheet modulus responds significantly, and the response can be

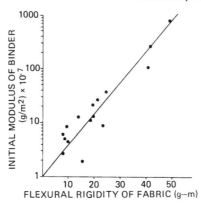

FIG. 14. Variation of the flexibility of polyester nonwoven fabrics with initial binder modulus.

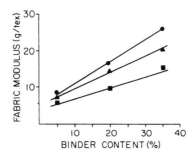

FIG. 15. Variation of fabric modulus with binder content at three levels of binder stiffness, 0. 2 (■), 0. 5 (▲) and 2. 0 (●) g/tex.

empirically expressed by the relationship

$$E_n = W \ (1.75 \log E_b + 0.69) \tag{26}$$

where W is the percentage binder content of modulus E_b (g/tex) and E_n (g/tex) is the corresponding modulus of the nonwoven.

However, although the selection of low modulus materials is a laudable theoretical approach to improve nonwoven drapeability, in practice the strength falls off quite rapidly with these softer binders. Inspection of examples of tensile failure in nonwovens bonded with these flexible polymers often reveals that the fiber is not breaking, but is pulling out of a sheath of polymer. A typical example of this phenomenon is provided by the behavior of a pair of viscose rayon fibers bonded with natural rubber, illustrated in Figure 16. Initially, the

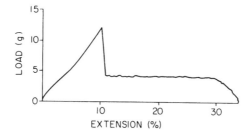

FIG. 16. Load-extension curve for two rayon fibers bonded together with natural rubber.

fibers stretch as the tension rises, but ultimately the fiber-rubber bond fails and then one fiber pulls through the collar-like bond with a constant frictional resistance until its extremity passes into the rubber and the tension begins to fall to zero. Actually, it is rather difficult to secure a good bond to a fiber without much manipulation because of the presence of surface contaminants originating from the fiber-manufacturing processes. In particular, low modulus adhesives which will stick well to rayon or other cellulosic fibers do not seem to have been discovered. The soft, rubbery materials now available are hydrocarbon in character and do not wet the cellulose well.

It may therefore be necessary to have some unambiguous bond between the fiber and the soft binder before the limit of performance for a fiber can be attained. To separate fiber and binder will then involve bond cleavage, and this could require as much as 80-100 kcal/mole for covalent bonds, which is equivalent to a load of about 10^{-8} g per bond. This concept in turn calls for fibers which either inherently or via chemical modification are capable of such bond formation. Of course, the bonding need not only be covalent in character, and binder-fiber chemical linkages spanning the entire bond-energy spectrum can readily be envisaged and will be discussed later in this chapter.

B. Binder-Fiber Interfacial Interactions

The majority of conventional adhesives now used in commercial nonwovens do not form discrete chemical bonds across the binder-fiber interface and the adhesion obtained must be attributed to some other type of interaction. Although numerous theories of adhesion have been proposed [30-34] to explain these interactions, it seems obvious from a chemistry point of view that ultimately the interactions must involve low-energy, intermolecular forces akin to the quasi-chemical bonds which are usually loosely described as van der Waals or London forces. Consequently, since bonds to fiber surfaces are involved, it seems equally obvious that the free energy of the surface must play an important role in determining the final nature of the adhesive joint. This thermodynamic concept can be of practical help in nonwoven bonding if it is borne in mind that the terms "surface free energy" and "surface tension" are interchangeable [35]. Perhaps the simplest method for the characterization of the state of a fiber surface, with respect to a particular binder, is by measurement of the contact angle θ. This technique [36] is not too demanding experimentally, and the results obtained are readily convertible into the work of adhesion W_{BF} using the classic Young-Dupré equation for low-energy surfaces where

$$W_{BF} = \gamma_B (1 + \cos \theta) \tag{27}$$

and γ_B is the surface tension of the fluid binder. Then, since the work of adhesion is a linear function of the strength of the resultant adhesive bond [37-39], this parameter can be used to assess and compare the potential utility of various fiber-binder combinations.

Clearly from Eq. (27) it follows that the work of adhesion will be maximized as the contact angle decreases, so that ultimately

$$W_{BF_{max}} = 2\gamma_B \tag{28}$$

when $\cos \theta = 1$. This is also the condition for maximum spreading of the binder on the fiber, and the spreading coefficient $S_{B/F}$ can be defined as

$$S_{B/F} = W_{BF} - W_{BB} \tag{29}$$

where W_{BB} is the work of cohesion of the binder, which by definition is also $2\gamma_B$. The larger the value of $S_{B/F}$, the more extensive the spreading, the larger the area of binder-fiber interface and, of course, the greater the extent of adhesion. Spreading can also be ameliorated by nonwoven processing variables. For example, since the surface tension of the binder, and hence the work of cohesion W_{BB}, diminishes linearly with increasing temperature [35, p. 55], hot pressing during nonwoven manufacture can be expected to improve bonding. Likewise, the addition to the binder of a small amount of suitable solvent will have a comparable effect, since the work of cohesion will be reduced as the magnitude of the interchain forces in the binder are diminished.

The usefulness of the spreading coefficients in practice is illustrated by comparison of the values calculated for the interaction of a

1:1 copolymer of isoamyl and neopentyl acrylate (γ_B, 26 dynes/cm) with polyethylene terephthalate ($S_{B/F}$, 19.4 ergs/cm^2), polycaprolactam ($S_{B/F}$, 16.4 ergs/cm^2), polystyrene ($S_{B/F}$, 6.2 ergs/cm^2), and polytetrafluoroethylene ($S_{B/F}$, -12.2 ergs/cm^2) [40].

From these data it can be concluded that this adhesive would be almost equally suitable as a binder for a polyester or polyamide nonwoven. Perhaps more usefully, it also can be inferred that polycaprolactam would be a reasonable choice for the binder in a rod- or spot-bonded polyester nonwoven, and indeed, this has been demonstrated [41].

It is therefore apparently implicit that the selection of the fiber determines the structure of the binder, since the latter must be capable of spreading on the former. However, such an implication need not necessarily be valid, because the chemical nature of the fiber surface can be modified by chemical treatments. This is exemplified by the data in Table 5, which shows the change in the contact angle of an epoxy-polyamide resin mixture on polyethylene which has been subjected to surface oxidation by chromic acid [42]. The drop in the contact angle with extended oxidation means that cos θ is increasing, and hence, from Eq. (27) it follows that the work of adhesion W_{BF} is also augmented. Since the cohesion of the binder is unaltered (γ_B, 41.7 dynes/cm) it is obvious from Eq. (29) that the value of the spreading coefficient must also have increased.

TABLE 5

Effect of Surface Treatment on the Contact Angle (θ) and Work of Adhesion (W_{BF}) of an Epoxy-Polyamide Resin[a] on Polyethylene

Surface treatment	Contact[a] angle deg	Work of[b] adhesion erg/cm^2
None	35.4	75.7
Acetone wipe (1)	33.8	76.4
(1) + 20 min in 23° chromic acid	26.5	79.0
(1) + 60 min in 23° chromic acid	22.0	80.4

[a]Epon 828-Versamid 140 (γ_B = 41.7 dynes/cm).

[b]Calculated from $W_{BF} = \gamma_B (1 + \cos \theta)$.

TABLE 6

Definition of the Work of Adhesion for Various Fiber-Binder
Combinations in Terms of Fiber-Binder Parameters

Binder	Binder parameters	Fiber	Fiber parameters	Work of adhesion
Nonpolar	$\gamma_B + \Delta_B = 0$	Nonpolar	$\gamma_F + \Delta_F = 0$	$2(\gamma_F \gamma_B)^{1/2}$
Nonpolar	$\gamma_B + \Delta_B = 0$	Polar	$\gamma_F + \Delta_F > 0$	$2(\gamma_F \gamma_B d_F)^{1/2}$
Polar	$\gamma_B + \Delta_B > 0$	Nonpolar	$\gamma_F + \Delta_F = 0$	$2(\gamma_F \gamma_B d_B)^{1/2}$
Polar	$\gamma_B + \Delta_B > 0$	Polar	$\gamma_F + \Delta_F > 0$	Eq. (30)

To deal with the work of adhesion in terms of the chemistry of the
modified fiber surface, it is necessary to employ the relationship developed by Good and coworkers [43-45] where

$$W_{BF} = 2(\gamma_F \gamma_B)^{1/2} [(d_F d_B)^{1/2} + (p_F p_B)^{1/2} + (\Delta_F \Delta_B)^{1/2}] \tag{30}$$

in which $d_F d_B$ and $p_F p_B$ are the dispersion and polar fractions of the
surface-energy density of the fiber and binder phases while Δ_F and Δ_B
account for the contribution to the work of adhesion of forces making up
hydrogen, ionic, or covalent bonds. Examination of Eq. (30) clearly
reveals the importance of compatibility between the binder and the fiber, since the largest value for the work of adhesion is attained only
when $d_B = d_F$ and $p_B = p_F$. In other words, the fractional molecu-
lar-force contributions to surface energy for both the fiber and the
binder should be in about equal proportions. This statement can be
illustrated by consideration of the work of adhesion associated with
the four discrete cases summarized in Table 6. Attention should be
specifically directed to the last combination, which affords the largest
work of adhesion because the term $\Delta_F \Delta_B$ does not disappear. Moreover, this term should be relatively large because of the higher energies associated with hydrogen bonding (1-10 kcal/mol), ionic interactions (10-30 kcal/mol), and covalent linkages (70-100 kcal/mol). The
potential magnitudes of the product $\Delta_F \Delta_B$ therefore provide a rationale

TABLE 7

Dispersion (γ_d) and Polar (γ_p) Contributions to the Surface
Tension (γ) of Polymers

Polymer	Dispersion contribution dynes/cm	Polar contribution dynes/cm	Surface tension dynes/cm
Polyhexamethylene adipamide	33. 56	7..77	41. 33
Polyethylene terephthalate	36. 59	2. 88	39. 48
Polyethylene	31. 29	1. 10	32. 39
Polyvinylidene chloride	38. 18	3. 16	41. 34
Polyvinyl chloride	38. 11	1. 51	39. 62

for research directed towards the development of discrete chemical
bonds across the fiber-binder interface.

Conversely, since the definitions of the work of adhesion for the
polar and nonpolar pairs each includes a dispersion factor (d_F or d_B)
which has a fractional value, it is plain that such binder-fiber combinations should probably be avoided. To underscore this point some
dispersion factor values have been calculated from literature data
[46] and are presented in Table 7. These, in turn, can be used to
predict the work of adhesion which hypothetically will be secured by
the fiber-binder combinations listed in Table 8. The implications of
the results in Table 8 are helpful in the confrontation of the practical
problems of bonding nonwovens comprising blends of fibers contrasting
chemical constitutions. Thus, the difficulty of bonding blends of rayon
and polyester fibers can be recognized in fundamental terms. Obviously, the laborious evaluation of an illimitable series of polymeric binders is not likely to be productive unless some novel method
of bridging the chemical gap between rayon and polyester can be con-
ceived.

1. Fiber Surface Rugosity

Although the chemistry of the interfaces largely determines the
degree of wetting, the topography of the fiber surface will also play
a role in the initial interfacial phenomenon as the binder makes contact. This surface effect, of course, has been known for a long time,
and the practice of abrading the adherend before application of the adhesive is well established. The abrasion increases the surface area,

TABLE 8

Theoretical Work of Adhesion for Some Fiber-Binder Combinations

Fiber	Binder	Work of adhesion erg/cm^2
Polyhexamethylene adipamide	Polyethylene terephthalate	79. 54
Polyhexamethylene adipamide	Polyethylene	70. 66
Polyethylene terephthalate	Polyvinylidene chloride	80. 80
Polyethylene terephthalate	Polyhexamethylene adipamide	79. 54
Polyethylene terephthalate	Polyethylene terephthalate	70. 94
Polyethylene terephthalate	Polyvinyl chloride	78. 84
Polyethylene terephthalate	Polyethylene	71. 24

reduces the contact angle of the adhesive, and thereby augments the work of adhesion. If the surface area is increased by the factor r then

$$r \cos \theta = \cos \theta_r \qquad\qquad (31)$$

where θ_r is the contact angle of the binder on the rough surface. It is therefore apparent that the interaction between binder and fiber can be improved by either physical or chemical treatment of the fiber.

These effects will often be intertwined, as illustrated by the example of delustered fibers. The delustering of synthetic fibers by the incorporation of inorganic particles increases the roughness [47, 48], while the chemistry of the fiber surface is also modified where the inorganic structures protrude sufficiently to be accessible to the binder.

IV. FIBER SURFACE MODIFICATION

Since the work of adhesion can be increased by modification of the chemistry of the fiber surface it is now appropriate to consider how this fundamental concept can be applied to the practical bonding of nonwovens. From the theory encapsulated in Eq. (30) it is apparent that the creation of unambiguous chemical bonds between binder and

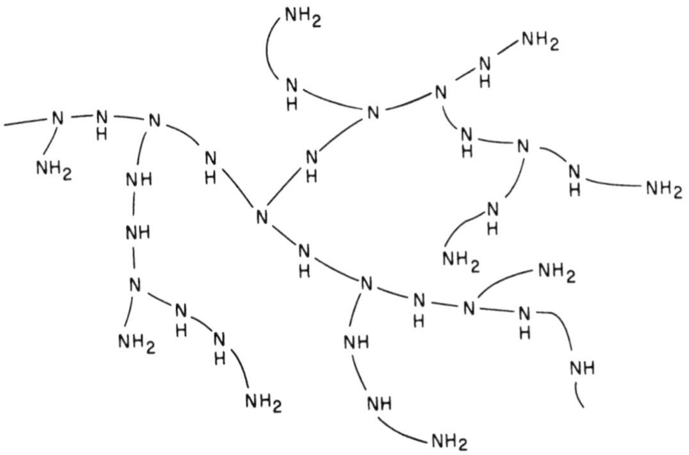

FIG. 17.   Idealized representation of the chemical constitution of polyethyleneimine.

fiber should be helpful in securing the desirable combination of tensile strength and drapeability in the nonwoven. Unfortunately, this is easier said than done, since many commercial fibers do not present any chemical site which can be easily modified. Polyester fibers are probably the most important of this group, and their reluctance to chemically interact with a wide range of molecules is manifested by the extreme difficulty of their dyeing [49]. These materials, however, like most other synthetic fibers, have a somewhat porous surface structure, the extent of which is very much dependent on the particular manufacturing process (see Chap. 4). These inhomogeneities offer an unusual opportunity for the utilization of the concept of polyelectrolyte entrapment developed about five years ago at the University of Washington [50]. This so-called Jack-in-the-Box process simply involves the initial application of a solution of the polyelectrolyte, in the condition of minimum hydrodynamic volume,to a microporous substrate. Subsequent variation in the character of the environment of the polyelectrolyte solution (usually pH) causes the polyelectrolyte to suddenly expand and become much larger in size. As a result of this expansion, macromolecules are trapped within containing pores of suitable size to provide a chemically modified surface.

Polyethyleneimine (PEI) is a very suitable polymer for this type of fiber-surface modification because it has a high charge density (Fig. 17), a large expansibility, and is commercially available in several molecular sizes. Application of PEI of various molecular weights to a polyester fiber using the Jack-in-the-Box procedure gave the retentions depicted in Figure 18 [51]. The amount of a particular

FIG. 18.   The effect of the size of polyethyleneimine on its retention by polyethylene terephthalate fibers (Celanese Corporation, T-400,
1.5 den., 6.4 mm staple).

size of PEI trapped is indicative of the number of pores or cracks of
this approximate dimension.   Clearly there is a wide size distribution
of these discontinuities.   In making measurements of this type, the
entrapment must be carried out under strictly specified conditions
because the equilibriation time as well as the initial and final pH of
the polyelectrolyte will all have an effect on the retention, as illustrated by the data in Figures 19 and 20.

A noteworthy feature of this method of fiber-surface modification
is that while the surface is made more polar, the surface rugosity is
essentially unchanged.   As yet, however, little is known of the mechanical strength of the entrapment of the polyelectrolyte inside the pores
or cracks.   If the polyelectrolyte acts only as an anchor holding the
binder onto the fiber, then the strength of the adhesive joint could be
limited by the strength of the polyelectrolyte-fiber mechanical link.
The strength of this link is certainly not negligible and in some experiments with bark fibers embedded in an epoxy resin, the PEI-fi-
ber linkage was strong enough so that the PEI-treated fiber broke
under strain, in contrast to a comparable but untreated fiber, which
merely slipped through the enveloping epoxy matrix.

A more visual example of the change in surface character of polyester fibers which can be achieved by the Jack-in-the-Box process is
provided by exposure of the PEI-treated fibers to a solution of a direct
dye such as Congo Red(I).   The fibers become strongly colored while

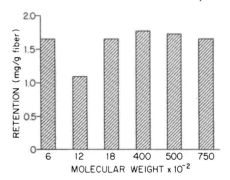

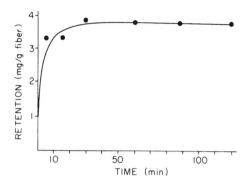

FIG. 19. The rate of polyethyleneimine (MW 1800) retention by polyethylene terephthalate fibers (E. I. DuPont de Nemours & Co., type 54W, 1.5 den., 6.4 mm staple).

untreated counterparts retained essentially no dye. This experiment demonstrates that the interaction of comparatively inert fibers to suitable chemical moieties can be drastically changed by simple means. The surfaces of the inert synthetic fibers are thus rendered similar, at least in part, to the more reactive fibers typified by rayon. Modification of these reactive fibers to introduce new bonding sites can be visualized at several distinct enthalpy levels of interaction ranging from van der Waal's forces upwards.

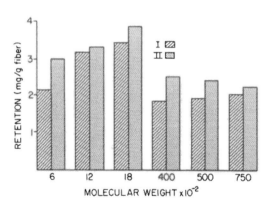

FIG. 20. The effect of the size of polyethyleneimine on its retention by polyethylene terephthalate fibers (E. I. DuPont de Nemours & Co., type 54W, 1.5 den., 6.4 mm staple) before (I) and after (II) acidification (to pH 5.0) and dilution prior to washing.

A. Hydrogen Bonding

Ascending the ladder of bond energies towards the upper limit of covalency, the next rung encountered is rather broad. Customarily categorized as hydrogen bonding, it encompasses enthalpies of from 1-10 kcal/mol.

Although a great deal of work has been devoted to hydrogen bonding, comparatively little is known about the stereochemical factors influencing the formation and strength of this bond category. In spite of the rather low enthalpies of hydrogen bonding, it is probably the most compatible with the extremely commercially significant cellulosic fibers because aggregates of the cellulose macromolecules are themselves held together by hydrogen bonds, and in paper, a special kind of nonwoven, these are also responsible for the cohesion of the sheet [4]. However, in spite of its key importance, detailed knowledge of the character of this hydrogen bonding is not extensive. Thus, infrared spectroscopy and also deuterium-exchange experiments indicate that in cellulose, almost all the hydroxyl groups (circa 96%) are hydrogen-bonded within the confines of the macromolecules, and only less than about 2% participate in interfiber bonding in the papermaking process [52]. In addition, Corte and Schaschek have pointed out [53] that while two distinct wave numbers for the vibration of free and bonded hydroxyl groups are found in simple molecules, cellulose shows a continuum of infrared absorption in the same range of wavelengths. Their interpretation of this spectrum is that cellulose is polymolecular with respect to the energy of its hydroxyl groups and that there is an energy range for the inter-fiber hydrogen bonds in paper which have an average value of 4.5 kcal/mol. It can therefore be inferred that cellulosic surfaces are capable of forming hydrogen bonds over a range of interatomic distances with practically any chemically receptive molecule. However, it must be emphasized that cellulose in actual use will always be associated with water molecules which will have filled all its potential hydrogen-bonding sites. New hydrogen bonds to the cellulose macromolecule itself can only be formed by displacement of this combined water by more stable ligands. Clearly, the stability of these new combinations will be determined both by the conformations of the cellulose and the ligands. The stereochemistry of the former is of course established and fixed as shown in Figure 21, while the shape and structure of the latter is susceptible to infinite variation. It should therefore be feasible to design molecules capable of fitting precisely onto the molecular contours of the cellulosic fiber surfaces so that more ordered interactions of relatively high enthalpy are secured.

Following such an approach, Allan and coworkers pointed out that the stereotopochemistry of α-cellulose fibers could readily be probed using nitrogen heterocycles because of the vast array of hydrogen-bonding conformations available within this group of compounds. It

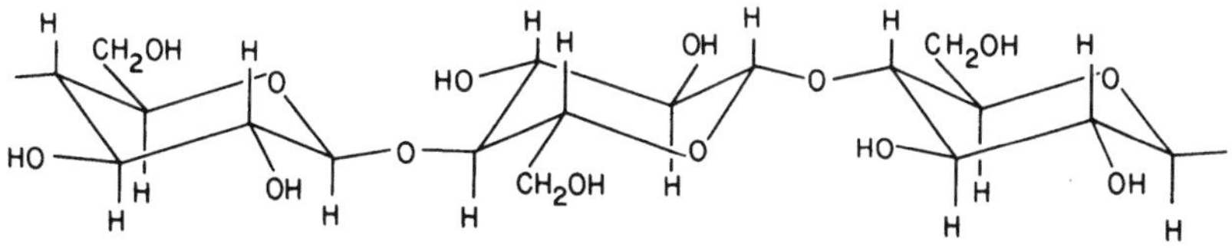

FIG. 21. Stereoformula of a segment of the cellulose chain.

was also suggested, because of nature's use of multiple hydrogen-bonding in the double helix of DNA [54] (Fig. 22), that the preferred molecular conformations to fit on α-cellulose might be found in the purine or pyrimidine clans. In an exploration of this idea a series of conveniently accessible 1-hydroxyimidazoles (IIa-l) structurally related to the nitrogeneous bases of DNA were equilibriated with α-cellulose fibers.

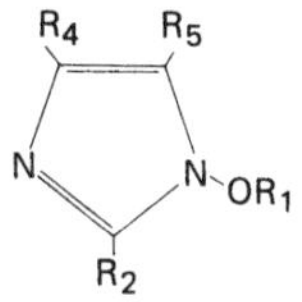

	R_1	R_2	R_4	R5
(IIa)	H	Me	Me	Me
(IIb)	H	Ph	Me	Me
(IIc)	H	pyrid-2-yl	Me	Me
(IId)	H	pyrid-3-yl	Me	Me
(IIe)	H	pyrid-4-yl	Me	Me
(IIf)	H	pyrazinyl	Me	Me
(IIg)	H	pyridazin-3-yl	Me	Me
(IIh)	H	pyrid-2-yl	pyrid-2-yl	pyrid-2-yl
(IIi)	CH_2=CHCO-	pyrid-2-yl	Me	Me
(IIj)	CH_2=CHCO-	pyrid-3-yl	Me	Me
(IIk)	CH_2=CHCO-	pyrid-4-yl	Me	Me
(IIl)	CH_2=CHCO-	pyrazinyl	Me	Me

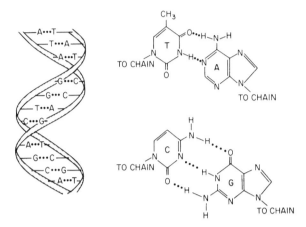

FIG. 22.   Double helix of DNA and the pairing of its constituents, adenine (A) - thymine (T); cystosine (C) - guanine (G).

The retentions found were significantly dependent on the spatial conformations of the constituent nitrogen atoms as shown in Figure 23. Of course, the retention of any substance by α-cellulose fibers is also affected by the pore-size distribution, which can be determined by equilibriation of the fibers with a homologous series of macromolecular sugars [55], ethers [56], or amines [57]. This size effect explains the lower receptivity of the larger (10Å) aromatic imidazole (IIb) relative to its smaller (8Å) aliphatic counterpart (IIa). However, the increased retention of all the pyridylimidazoles (IIc-e) compared to that of the equally large phenylimidazole (IIb) is due to the presence of the additional nitrogen atom which provides another bonding linkage to the fiber. The difference in receptivity between the isomeric 3- and 4-pyridyl derivatives reflects the statistical geometric advantage of the former (IId), which can achieve the most suitable spatial conformation on the fiber surface by virtue of the free rotation about the bond joining the pyridine and imidazole rings. The 3-pyridylimidazole should thus be able to reach more bonded sites than its 4-pyridyl isomer (IIe) where the rotation does not change the spatial arrangement of the pyridyl nitrogen. The lowest receptivity was observed with the 2-pyridyl analog (IIc), which most likely is due to the intervention of intramolecular hydrogen bonding which diminishes the opportunity for the formation of intermolecular bonds between the imidazole and the hydroxyl groups of the fiber.

The benefits of introducing additional nitrogen atoms as in IIg and IIh, although significant, are not dramatic. The excellent retention of the pyrazinylimidazole (IIf) with its four nitrogen atoms located comparably close to those in the purines is supportive of the preferred conformation concept outlined in the foregoing.

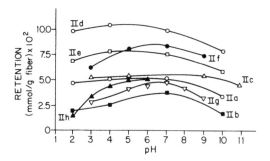

FIG. 23.   The effect of the pH of equilibration on the retention of 1-hydroxyimidazoles by α-cellulose.

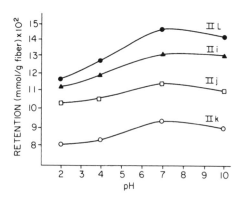

FIG. 24.   The effect of the pH of equilibration on the retention of 1-acryloxyimidazoles by α-cellulose.

To use these findings in the design of improved nonwoven binders requires incorporation of the imidazole moiety into polymers, and this can be most conveniently realized through the acrylate esters of the 1-hydroxyimidazoles (IIi-l).   However, the esterification of the 1-hydroxyl group destroys the amphoteric character [58] and generates a new pattern of bonding to the α-cellulose fibers, which is illustrated by the receptivity data in Figure 24.   The retention of the 2-pyridyl-imidazole acrylate is now superior to its 3- and 4-pyridyl isomers. This must reflect its ability to secure a better fit upon the polysaccharide conformation of the cellulose surfaces.   The construction of Dreiding stereomodels showed that imidazole derivatives of this type can theoretically form hydrogen bonds of the plug-and-socket type, as illustrated in Figure 25 by the excellent fit on the α-cellulose structure.   All these results serve to show the importance of the spatial arrangement of bonding sites within individual molecules as

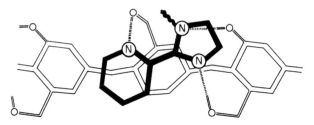

FIG. 25.   Drawing of Dreiding stereomodels showing fit of 1-sub-
stituted 2-(pyrid-2-yl)imidazole (black) on cellulose units (outlined)
and hydrogen bonding (dashed) between nitrogen atoms of adsorbate
and hydroxyl groups of cellulose.

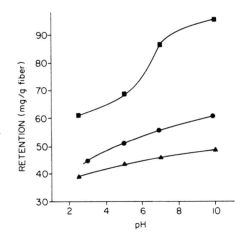

FIG. 26.   The effect of the pH of equilibration on the retention of
pyrid-2- (■), -3- (●) and -4- (▲) ylimidazolic polyacrylates by α-
cellulose.

a means of securing the highest possible interaction of binder with the
surfaces of the fiber. However, to be useful in binder design these enhanced levels of interaction of monomers must carry over to polymers,
and this is indeed the case.   Thus, the corresponding series of polyacrylates at comparable molecular weights also exhibit distinctly different degrees of adsorption, which are presented in Figure 26 and
which parallel the retention behavior of the monomers. Although the
merit of inclusion of these form-fitting moieties in low-modulus binders would seem to be clear, nonetheless, the minimum effective level
of incorporation remains to be established.   It can be expected that this
level will vary somewhat depending on the extent of the interaction of

TABLE 9

The Retention of Bolas Polymers and Their Progenitors
by α-Cellulose Fibers •

Polymer	Molecular weight	Retention (mg/g fibers)
Polyoxypropylenediamine	1000	1.59
Polyethyleneglycol	2000	
$\alpha\omega$-Di(2, 4, 5-tripyrid-2-ylimidazo-1-yl)-amino-polyoxypropylene	1600	32.0
$\alpha\omega$-Di(1-phenyl-1H-tetrazo-1-yl)-amino polyoxypropylene	1300	20.7
$\alpha\omega$-Di(3-sodium sulfonate-4-methoxyphenylenethylene)-amino polyoxypropylene	1410	17.8
$\alpha\omega$-Di(2, 4)-dipolyethyleneiminyl-s-triazinyl-6-oxy)polyoxyethylene	5000-7000	47.7

the other comonomer units in the binder with the fiber surface. Even
a very low level of incorporation can have a significant effect, as is
illustrated by the almost twenty-fold increase in adsorption of poly-
oxypropylenediamine (MW, 1000) as a consequence of the attachment
of a tripyrid-2-ylimidazole moiety at each end of the polymer chain
(entries 1 and 3, column 3, in Table 9). A similar effect was observed using the polymer obtained by the reaction of each of the end-
groups of the polymeric diamine with 5-chloro-1-phenyl-1H-tetrazole
[59] (entry 4, Table 9). These increases in retention are not entirely due to the chemistry of the polymers but are, in part, a reflection of their peculiar geometry, which makes some contribution.
That is, the polymer can be pictured as consisting of two bulky spher-
oids linked by a long line. This is reminiscent of the formidable weap-
on of the Argentinian gaucho [60], which consists of balls of stone attached to the ends of a rope of twisted or braided hide or hemp. When
accurately thrown, the whirling balls wind around the target, cross
each other and become firmly hitched. Presumably this form of entanglement can occur on the surface of some fibers, and these macromolecular structures can therefore be aptly termed bola polymers.
Some preliminary confirmation of this quaint mechanism of binder-
fiber interaction was provided by a study of the retention of other bola

614 ALLAN, LAINE, AND NEOGI

polymers on fibers. For example, the Schiff's base formed by the reaction of 2-methoxybenzaldehyde-4-sulfonic acid with the same polyoxypropylene diamine gave a polymer with negative sites at the extremities. In spite of the repelling effect of these sites [61] towards adsorption on cellulose, the retention of this bola polymer by intrinsically negative α-cellulose fibers was more than eleven times that of its unmodified, positively charged progenitor, diamine (entry 5, Table 9). An even more spectacular illustration of the effect of this type of binder geometry on the intensity of the binder-fiber interactions was provided by a study of the retention of the polymer synthesized by attachment of four spherical polyethyleneimine macromolecules (MW, 1800) to the extremities of a polyoxyethyleneglycol (MW, 2000) using cyanuric chloride as a bridging entity between the polymeric components. Of course, in this case the bola effect cannot be completely disentangled from the contributions which originate from the polyionic character of the bulky ends, which separately have a significant amount of independent ionic interaction with cellulosic fibers [62]. Obviously, in the future the role of binder molecular geometry in binder-fiber interactions will be the subject of increasing scrutiny, and it seems likely that among the vast array of shapes, such as combs and stars [63], polymers will be found to replace the essentially linear macromolecules now used.

B. Ionic Bonding

Continuing up the ladder of bond energies, ionic fiber-binder interactions constitute the next category for consideration. Ionic bonds are of course the essential cement of inorganic chemistry, and their enthalpies range from 10 to 30 kcal/mole. In spite of the high cohesive strength of many inorganic materials comparatively little attention has been given to the utilization of this type of bond in fiber-polymer composites, although Nissan and Sternstein [64] and Page [14] have shown how the ultimate strength of the nonwoven paper is determined by the strength and number of bonds connecting contiguous fibers. Obviously, to construct an ionically bonded nonwoven requires that both the fiber and binder have ionic sites and that these are opposite in charge. In practice it usually is somewhat easier to introduce anionic sites onto a fiber, and indeed, commercial rayon already contains a number of such groups formed by the oxidation of the precursor cellulose pulp during the bleaching process. Negative sites can also be created on the surfaces of nylon, polyester, and polyacrylonitrile fibers by hydrolysis. Thus far there is not yet commercially available a wide selection of binders containing positive charges. The polyethyleneimine family, ranging in molecular weight from 1800 to 75000, is perhaps the best known. However, the amino polysaccharide, chitosan(III), is also beginning to be available commercially, and will undoubtedly play an important role in the future

III

because of its outstanding film-forming properties. Some cationic poly-acrylamides are also sold as flocculants for water treatment but their use as nonwoven binders has been only incidental [65]. On the other hand, there are several cationic vinyl monomers which are commer-cially available including the vinyl pyridines, as well as dimethyl-aminoethyl and t-butylaminoethyl methacrylate [66].

These can be conventionally copolymerized with the standard bind-er monomers to afford a range of cationic binders. However, in the studies of nonwovens which has been reported, polyethyleneimines thus far have been the binders of choice because of their convenient avail-ability in a wide range of molecular sizes. This is exemplified by the bonding of a preformed rayon with a spray of polyethyleneimine (MW, 75000) solution at pH 7. The retention of PEI by the web, as evidenced by the data in Table 10, was essentially constant and unrelated to the amount sprayed. This emphasizes a most important peculiarity of ionic bonding, which is the desirability of achieving an ionic balance be-tween the binder and the fiber. Since the number of advantageous ionic sites on the fiber is normally not large, a logical approach to building more energetic binder-fiber interfaces is to increase the number of such sites. This can be smoothly accomplished, especially for rayon nonwovens, by using fiber reactive dyes [67]. By the sequence shown in Structure IV the dye becomes covalently bonded to the fiber and its constituent anionic sulfonate sites henceforth are an integral part of

IV

TABLE 10

Polyethyleneimine (PEI) Retention and the
Tensile Strength of Rayon Webs

PEI addition mg/g	PEI retention mg/g	Breaking length, m
0	9	450
60	4.9	917
120	4.5	1038
240	5.6	1002

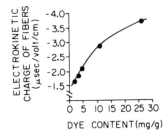

FIG. 27. Variation of the anionic character of α-cellulose fiber
with dye content. Reprinted from Ref. 67, by courtesy of Textile Research Institute.

the fiber. Up to 6% by weight of the modifier can easily be incorporated
and this is equivalent to some 10^{20} anionic sites per gram of fibers.
However, all of these locales are not available for interfiber bonding.
This is illustrated by measurement of the electrokinetic charge of dyed
α-cellulose fibers. The data in Figure 27 shows that the potential does
not increase linearly with the anionic site content, and this deviation
serves as a measure of the inaccessible interior surface charges. This
inaccessibility is a consequence of the microporous structure of the fiber which Stone et al. [55] have characterized, using as probes a series
of spherical polysaccharides of different molecular diameters. Although
these sites are inaccessible for electrokinetic purposes, small poly-
amines can reach them and form intrafiber crosslinks which modestly
increase the absolute strength of the fiber (Fig. 28). Although dyeing
is a convenient method of introducing negative charges within fibers,
its value may be somewhat academic, since such obvious disadvantages

TABLE 11

Polyethyleneimine (PEI) Retention and Carboxylic Acid Content of Fiber Blends of Regular and Carboxymethylated Cellulose Rayon

CMC rayon (%)	0	20	40	60	80	100
COOH content mol/g fibers X 10^5	3.8	12.2	20.7	29.1	37.6	46.0
PEI retention mg/g fibers	4.2	8.0	9.6	11.0	12.0	12.3

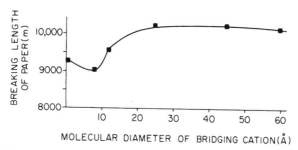

FIG. 28.   Variation of the zero-span tensile strength of ionically bonded paper with the size of the ionic bridge.

as the additional dying step, wasted internal dye, and the color conferred on the fiber possibly militate against its use in practice. Therefore, surface modification by alternate procedures are of interest.

Obviously, there are many alternative methods of fiber modification, and by virtue of its method of production from a solution of cellulose xanthate, rayon is particularly amenable to modification by co-spinning. Thus, if carboxymethyl cellulose (CMC) at various levels is dissolved in the xanthate pulp at various levels, the resultant CMC-rayon will contain a multitude of anionic carboxylic sites originating from the entangled CMC macromolecule [68]. A comparison of this CMC-rayon, regular rayon, and blends thereof in a series of nonwovens have shown that the retention of PEI (Table 11) led to an increase in the tensile strength (Fig. 29) which was approximately directly proportional to the amount of polyamine retained. The contribution of ionic bonds is even more outstanding when the wet strength of the nonwoven is considered. The magnitude of this property

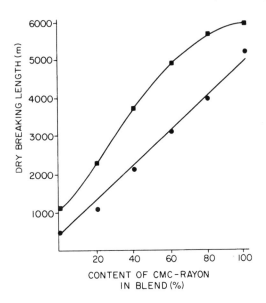

FIG. 29. Tensile properties of nonwoven blends of regular and CMC-rayon with (■) and without (●) application of PEI.

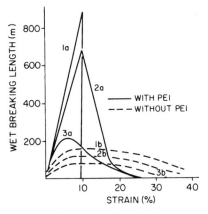

FIG. 30. The wet tensile strength-strain curves of nonwoven blends from regular and CMC-rayon; 1a-b (100% CMC-rayon); 2a-b (60% CMC-rayon); and 3a-b (20% CMC-rayon).

increased from 3 to 17% of the dry strength. Actually this is about the highest wet strength attainable with the CMC-rayon because almost all of the fibers across the failure line were ruptured. This type of failure is illustrated by some of the stress-strain curves in Figure 30.

The flat curves represent progressive cleavage of interfiber bonds with increasing strain.   In contrast, the curves with the abrupt fall-off reflect the progressive rupture of fibers as the strain is increased. This is the only form of failure in the 100% CMC-rayon nonwoven.

The benefits of such ionic bonding in rayon nonwovens can also be realized without the production of special rayon fibers, which is likely to be inherently costly because of the volume of production. For example, impregnation of regular rayon with an aqueous solution of a polymeric acid followed by acidification will yield a highly ionic rayon [69]. The marine polymer, alginic acid (V) is a particularly

ALGINIC ACID WITH D-MANNURONIC ACID UNITS

OR WITH SOME
L-GULURONIC
ACID UNITS:

V

suitable modifying material because of its high molecular weight, which restricts wasteful penetration, its polysaccharide chemical structure, which can readily form multiple hydrogen bonds to the cellulose of the rayon fiber, and its uniform distribution of anionic sites. Although the alginic acid itself acts as a binder, the data in Figure 31 is illustrative of the benefits of this approach to ionic bonding.

The advantages of alginic acid are further underscored by examination of the potential of rayon fibers grafted with acrylic acid.   Apart from the questionable economics of grafting, it is apparent from the data in Table 12 that the anionic-group content of grafted rayon can be much higher than that of CMC-rayon. As a result, the amount of PEI retained is also greater.   However, this increased retention was of relatively little benefit [70] and the wet strength of nonwovens [69] was far below that obtained with the CMC-rayon. It must therefore be concluded that the acrylic acid-grafted product has features which are not favorable to the construction of effective interfiber bonds. Most probably grafting will result in only a few long side chains of poly-acrylic acid too far apart from one another [71] to maximize inter-fiber interactions. The advantage of having the ionic sites quite uniformly spaced and accessible is evidently of sufficient importance to mandate the selection of the method for fiber-surface modification. Of course, it is also possible to combine the beneficial attributes of ionic bonding with those of the lower energy levels. The polymer,

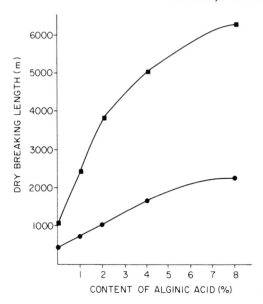

FIG. 31. Breaking length of alginic acid-treated rayon nonwoven with (■) and without (●) application of PEI.

TABLE 12

Polyethyleneimine (PEI) Retention and Carboxylic Acid Content of Fiber Blends of Regular and Acrylic Acid Grafted Rayon

Grafted rayon	0	20	40	60	80	100
COOH content mol/g fibers $\times 10^5$	3.8	27.C	50.3	73.5	96.8	120
PEI retention mg/g fibers	4.2	7.5	16.4	25.8	34.3	41.8

chitosan, is a particularly good example of this. Chitosan (III) can be regarded as an acid-soluble cellulose, and as such can form multiple hydrogen bonds very readily within rayon nonwovens. In addition, the regular distribution of basic amino groups along the large linear mac-romolecule means that the probability of an anionic site failing to con-tact a cationic center within the binder is negligibly low. The film-forming propensities and the film strength of chitosan are also of sub-stantial benefit in its role as a nonwoven binder.

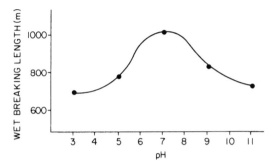

FIG. 32. Wet breaking length at various pH values of PEI treated CMC-rayon nonwoven.

Finally, whatever the mode of ionic-site introduction, it is important to realize the critical effect of pH on bond formation. It is clear that the combination of macroions to form bonds will be sensitive to the presence of interfering ionic species, which of course include hydrogen and hydroxyl ions. Illustrative of this point is the variation in the wet strength of CMC-rayon nonwovens, shown in Figure 32, wet-formed at different pH values. A definite maximum is observed and the location of this point is determined by the pK_a values of the interaction ionic species.

C. Covalent Bonding

The top of the bond-enthalpy ladder corresponds to the ultimate bonding situation where the binder-fiber interface is spanned by covalent bonds. These linkages are much higher in energy than the other types of bonds and are directly comparable in strength to the backbone bonds of the fiber macromolecules themselves. That is, the force to separate the binder interface should be similar to that needed to break the fiber. The attempted formation of covalent linkages between binder and fiber is not new, and it has long been suspected that the classic wood adhesives based on phenolformaldehyde resins reacted with the substrate. Recently, using the brominated benzyl alcohol (VI) as a model, unequivocal evidence has been presented [72] which demonstrates that methylol groups attached to aromatic rings are capable of extensive reaction with lignocellulosic but not cellulosic fibers. Likewise, the advantages of covalent binder-fiber linkages are well established in the area of fiber-glass composites, and modification of the glass-fiber surface with silicon derivatives containing vinyl and amino or glycidyl substituents improves the adhesion to unsaturated polyesters and epoxy resins respectively [73]. However, the idea of building on this basic background to develop covalent bonding systems seems to have been inhibited by the belief that extreme conditions are required to achieve reactions at functional sites on fiber surfaces. This attitude is typified by research done on mercerized cellulose in highly alkaline media [74]. The vast body of literature on the many types of fiber-reactive dyes which combine with both natural and synthetic fibers under mild conditions shows how misguided this view can be.

Of the several chemicals used to create the fiber-reactive moiety in these dyes, cyanuric chloride is certainly the most available commercially, and binders containing fiber-reactive units derived from this versatile chemical will probably appear in the future. Already it has been demonstrated that pulp fibers can be modified using cyanuric chloride so that the surfaces have colorless pendant chloro-s-triazinyl groups [75]. These sites are later capable of being crosslinked using diamines, to afford papers which will maintain their coherency in strongly agitated water and even resist dissolution in a cellulose solvent such as Cadoxen, a cadmium ethylenediamine hydroxide complex [76, 77].

The protection of cellulosic fiber networks from the disintegrating effect of water is of course a problem solved long ago by Nature. Thus, a tree is a mass of cellulose fibers held together by the binder, lignin. This binder covalently links the fiber very effectively under the mildest of conditions. The genesis of this binder is generally accepted [78] to be a consequence of a complex enzymic dehydrogenative copolymerization of mixtures of various ratios of coniferyl (VIIa), sinapyl (VIIb), and p-coumaryl (VIIc) alcohols through the intermediacy of the derived free radicals (VIIIa-e).

These alcohols are all basically p-vinylphenols, and because of the mode of polymerization, which affords the binder lignin, other compounds containing a replaceable hydrogen and contiguous to the locus of lignification may also be incorporated into the macromolecule by chain transfer, radical coupling, or addition across the chromophore of the quinone methides derived by rearrangement from the oxidized p-vinylphenolate anions.

Clearly this reaction is readily adaptable to the bonding of cellulose nonwovens, especially in pulp fibers, where the residual lignin can be regarded as a cataleptic polymer awaiting the appropriate external stimulus [86]. This is provided by treatment with an oxidizing agent

(VIIIa) R_1 = OMe; R_2 = H; R_3 = CH_2OH

(VIIIb) R_1 = R_2 = OMe; R_3 = CH_2OH

(VIIIc) R_1 = R_2 = H; R_3 = CH_2OH

(VIIId) R_1 = OMe; R_2 = H; R_3 = CH_3

which converts the phenolate anions into the corresponding radicals which are effectively immobilized and isolated within the lignin matrix. The existence of such stabilized radicals in wood and lignin has been known for some time [79]. The activated fibers are then receptive to combination with other free radicals [80-82].

In the paper area, this opportunity to create covalent interfiber linkages has been seized and used as a springboard to utilize lignin-derived pulp wastes. That is, treatment of both unbleached kraft and groundwood assemblages with the waste lignins and oxidizing agents causes the latter to covalently combine with the former. Since there are practically no absolute termination reactions the grafting of the lignin to the fiber can be continued indefinitely. By subjecting these fiber structures to a number of repetitive grafting cycles it can be shown (Figs. 33, 34) that the tensile strength after each cycle increases and finally passes through a maximum [83]. The tensile behavior of these two fiber networks indicates that a coherent,

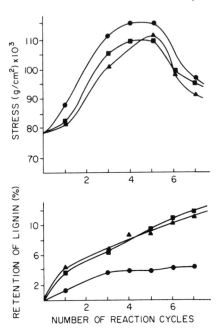

FIG. 33. Lignin retention and stress values (load to break/cross-sectional area of a test strip of 15 mm in width) for unbleached kraft fiber handsheets treated with kraft lignin (●), ammonium lignin sulfonate (■), or sodium lignin sulfonate (▲), and potassium ferricyanide.

covalently bonded structure is being built up, the adhesive points of which are able to transmit higher stresses from the binder to the fibers without disruption. The ultimate drop in tensile strength observed is probably a consequence of the formation of an excessive amount of binder which is then susceptible to cracking-stress localization and early failure. However, in principle, this method of covalent bonding of fibers should not readily carry over to nonwovens made from lignin-free rayon because chain transfer to or radical coupling with cellulose are of rare occurrence due to the high redox potential of cellulose and the low stability of the few cellulose radicals formed. Nonetheless, among the several modes of combination of the free radicals derived from p-vinylphenols, those involving the quinone methide radicals (VIIId) can readily react with polysaccharide alcohols. These functional groups are therefore potentially

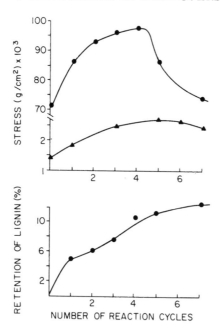

FIG. 34. Lignin retention and wet (▲) and dry (●) stress values (load to break/cross-sectional area of a test strip of 15 mm in width) for unbleached groundwood treated with sodium lignin sulfonate and potassium ferricyanide.

suitable for the construction of covalent linkages to rayon fibers which are replete with alcoholic hydroxyl groups. The competitive addition of water across the quinone methide moiety can be reduced by application of the p-vinylphenol precursor to the fiber phase before dehydrogenation is initiated.

The feasibility of this covalent bonding approach to rayon has been demonstrated by Allan, Bullock, and Neogi [84], employing isoeugenol (VIIId) as a conveniently available p-vinylphenol. Using aligned bundles of hollow rayon fibers impregnated with isoeugenol, enzymic oxidation induced a synthetic counterpart of lignification. The covalently bonded structures obtained had physical strength properties comparable to those of commercial U.S. softwood lumber.

The generality of this mild reaction system for bonding to rayon was further established by the development of a new fiber-reactive dye system. Thus, diazotization of 4-amino-1-naphthalene sulfonic acid (IXa) and coupling with isoeugenol yielded a dye (IXb) containing a p-vinylphenol moiety. Application of this dye to purified cotton fibers followed by oxidation resulted in the formation of covalent bonds between the dye and the fibers [85]. Undoubtedly this type of reaction on fiber surfaces presages the evolution of new kinds of nonwoven binders.

IX a

IX b

V. CONCLUSIONS

All the considerations reviewed in this chapter should make pellucidly evident the fact that nonwovens have largely evolved from either the pre-existent textile or paper industries. At present, nonwovens use the materials specifically developed for these mature enterprises.

As a consequence, many of the apparent constraints faced by the fledgling nonwoven industry are fundamentally irrelevant and have their origins in the economics and practices of the parent textile and paper concerns.

It can therefore be anticipated that as the fiber and binder manufacturers increasingly come to regard nonwovens as an important market entity in its own right, rather than as an adjunct opportunity, products specifically tailored for nonwovens will be widely offered.

This already occurring change in attitude is being accompanied by the development of new mechanical methods of producing and handling films, fibers, binders, and webs so that progress in the next decade is likely to be extremely rapid.

Nevertheless, whatever forms of nonwovens are evolved in the future, analytical application of the concepts presented here will help rationally establish the best combination of surface properties.

GLOSSARY OF SYMBOLS

T	tensile strength
f	parameter defined by the content and distribution of binder
Z	zero-span tensile strength
f_p	fraction of fibers pulled out across the line of failure
σ_f	fiber tenacity
σ_b	bond strength per unit bonded length of fiber
ℓ_b	average bonded length per fiber
B	index of bonding
n_f	number of fiber centers per unit area
σ_a	shear strength of adhesion
P_a	perimeter of adhesion
K	constant of proportionality
P_f	perimeter of the fiber
A	cross-sectional area of the fiber
n	number of sides of noncircular fiber
I	moment of inertia
P_{cr}	critical pressure for circular fiber collapse
E	modulus of elasticity
h	fiber wall thickness
d	arithmetic mean of the core and exterior diameters of a hollow fiber

d_c core diameter of a hollow circular fiber

d_s exterior diameter of a hollow circular fiber

P_s critical pressure for hollow square fiber collapse

P_t critical pressure for hollow triangular fiber collapse

L_d exterior dimension of hollow square and triangular fibers

ℓ_d interior dimension of hollow square and triangular fibers

n_t number of fibers enveloped by a binder spot

r_b radius of a circular binder spot

a ratio of fiber length to binder spot diameter

ℓ fiber length

k constant of proportionality

N total number of bonding sites

n_b number of binder spots per unit area

E_n elasticity of nonwoven, g/denier

W percentage binder content

E_b elasticity of binder, g/denier

θ contact angle between binder and fiber

W_{BF} work of adhesion between binder and fiber

γ_B surface tension of binder

W_{BB} work of cohesion of binder

$S_{B/F}$ spreading coefficient of binder on fiber

γ_F surface tension of fiber

d_F, d_B dispersion fractions of the surface energy density of the fiber and binder, respectively

p_F, p_B polar fractions of the surface energy density of the fiber and binder, respectively

Δ_F, Δ_B total interacting forces less polar (p) and dispersion (d) contributions of the fiber and binder, respectively

r roughness factor of substrate

θ_r contact angle of binder and a rough fiber surface

REFERENCES

1. G. G. Allan, in "Theory and Design of Wood and Fiber Composite Materials," (B. A. Jayne, ed.), Syracuse University, Syracuse, N. Y., 1972, pp. 299-326.
2. G. G. Allan and A. N. Neogi, Cellul. Chem. Technol., 8, 141 (1974).
3. H. F. Arledter, in "Synthetic Fibers in Papermaking," (O. A. Battista, ed.), Interscience, New York, 1964, p. 10.
4. O. A. Kallmes, in "Theory and Design of Wood and Fiber Composite Materials," (B. A. Jayne, ed.), Syracuse University, Syracuse, N. Y., 1972, pp. 157-175.
5. R. L. Adelman, G. G. Allan, and H. K. Sinclair, Ind. Chem. Eng. Chem. Prod. Res. Devel., 2, 108 (1963).
6. A. H. Drelich and H. W. Griswold, U.S. Pat. 2,880,111 (1959); U.S. Pat. 3,009,822 (1961).
7. A. H. Drelich, U.S. Pat. 2,880,112 (1959); U.S. Pat. 2,880,113 (1959); U.S. Pat. 3,009,823 (1961).
8. G. G. Allan and G. D. Crosby, Tappi, 51, 92A (1968).
9. G. G. Allan, B. K. Garg, and M. L. Miller, Tappi, 54, 406 (1971).
10. F. Kalwaites, U.S. Pat. 3,681,184 (1972); U.S. Pat. 3,769,659 (1972).
11. E. M. Passot, in "Computer Evaluation of Fiber Crimp and Spot-Bonding for Drapeable Nonwovens," Ph.D. Thesis, University of Washington, 1974.
12. W. F. Cowan, Pulp Pap. Can., 46, 86 (1972).
13. J. W. S. Hearle, in "Theory and Design of Wood and Fiber Composite Materials," (B. A. Jayne, ed.), Syracuse University, Syracuse, N. Y., 1972, pp. 327-351.
14. D. H. Page, Tappi, 52, 674 (1969).
15. J. d'A. Clark, Tappi, 56, 122 (1973).
16. H. L. Cox, Brit. J. Appl. Phys., 3, 72 (1952).
17. G. G. Allan and L. A. Smith, Cellul. Chem. Technol., 2, 80 (1968).
18. G. G. Allan, M. L. Miller, and A. N. Neogi, Cellul. Chem. Technol., 4, 567 (1970).
19. J. W. S. Hearle, R. I. C. Michie, and P. J. Stevenson, Text. Res. J., 34, 275 (1964).
20. R. A. A. Hentschel, in "Synthetic Fibers in Papermaking," (O. A. Battista, ed.), Interscience, New York, 1964, p. 290.
21. L. J. Broutman and R. H. Krock, in "Modern Composite Materials," Wesley, Reading, Mass., 1967, pp. 332-333.
22. American Enka Company, Tech. Bull. PFS-4, 1973.

23. Photomicrographs were kindly supplied by Dr. D. G. Bannerman of the Textile Fibers Department of E. I. du Pont de Nemours and Company, Inc., Wilmington, Del.

24. D. W. van Krevelen, Chem. Ind. (London), 1396 (1971).

25. K. Shimoda and K. Ban, Jap. Pat. 3,353,026 (1973).

26. G. G. Allan and W. McConnell, unpublished results, 1974.

27. R. O. Runkel, Das Papier, 3, 476 (1949).

28. A. A. Robertson and S. G. Mason, in "The Formation and Structure of Paper," (F. M. Bolam, ed.), Vol. 2, Technical Section of the British Paper and Board Makers' Association, London, 1962, pp. 639-650.

29. J. W. S. Hearle and A. Newton, Text. Res. J., 37, 495 (1967).

30. A. A. Berlin and V. E. Basin, Osn. Adgezii Polim. (1969) (Khimiya, Moscow).

31. J. J. Bikerman, in "The Science of Adhesive Joints," Academic, New York, 1968.

32. J. J. Bikerman, Usp. Khim., 71, 1431 (1972).

33. J. J. Bikerman, Vysokomol. Soedin., Ser. A, 10, 974 (1968).

34. D. H. Kaelble, in "Physical Chemistry of Adhesion," Inter-science, New York, 1971.

35. A. W. Adamson, in "Physical Chemistry of Surfaces," Inter-science, New York, 1967, p. 64.

36. T. H. Grindstaff, Text. Res. J., 39, 958 (1969).

37. M. J. Barbarisi, Nature, 215, 383 (1967).

38. M. Levine, G. Ilkka, and P. Weiss, Polymer Letters, 2, 915 (1964).

39. C. A. Dahlquist, ASTM Spec. Techn. Pub., 360, 46 (1963).

40. D. H. Kaelble, J. Adhesion, 1, 102 (1969).

41. C-L. Ong, M. S. Thesis, University of Washington, 1974.

42. M. J. Barbarisi, in "Wetting of High and Low Energy Surfaces by Liquid Adhesives and Its Relation to Bond Strength," Techn. Rept. 3456, Picatinny Arsenal, Dover, N. J.

43. R. J. Good, L. A. Girifalco, and G. Kraus, J. Phys. Chem., 62, 1418 (1958).

44. R. J. Good, in Advances in Chemistry Series, No. 43, p. 75, American Chemical Society, Washington, D. C., 1964.

45. R. J. Good, in "Treatise on Adhesion and Cohesion," (R. L. Patrick, ed.), Dekker, New York, 1967.

46. D. H. Kaelble, in "Physical Chemistry of Adhesion," Inter-science, 1971, pp. 164-165.

47. R. D. van Veld, G. Morris, and H. R. Billica, J. Appl. Polym. Sci., 12, 2709 (1968).

48. W. J. Lyons and M. G. Scott, Text. Res. J., 36, 585 (1966).

49. R. W. Moncrieff, in "Man-Made Fibers," Heywood, London, 1969, p. 372-377.

50. G. G. Allan, K. Akagane, A. N. Neogi, W. M. Reif, and T. Mattila, Nature, 225, 175 (1970).

51. J. E. Laine, K. Akagane, and G. G. Allan, in "Polyester Surface Morphology and Bonding in Paper Synthetics," Paper Synthetics Conference, Tappi, St. Louis, Mo., October, 1973.

52. H. Corté, in "Composite Materials," (L. Holliday, ed.), Elsevier, Amsterdam, 1965, p. 490.

53. H. Corté and H. Schaschek, Das Papier, 9, 519 (1955).

54. A. L. Lehninger, in "Biochemistry," Worth Publishers, Inc., 1970, pp. 640-641.

55. J. E. Stone and A. M. Scallan, Cellul. Chem. Technol., 2, 343 (1968).

56. J. A. Ciriacks, private communication, 1969.

57. G. G. Allan and W. M. Reif, Svensk Papperstidn., 74, 563 (1971).

58. F. J. Allan and G. G. Allan, Chem. Ind. (London), 1937 (1964).

59. K. Akagane, in "Fiber-Polymer Interactions," Ph. D. Thesis, University of Washington, 1972, p. 59.

60. C. R. Darwin, in "Charles Darwin and the Voyage of the Beagle," (N. Barlow, ed.), Philosophical Library, New York, 1946.

61. T. Mattila, in "Chemical Surface Modification of Lignocellulosic Fibers Using Cyanuric Chloride," Ph. D. Thesis, University of Washington, 1970, p. 91.

62. G. G. Allan and W. M. Reif, Svensk Papperstidn. 74, 25 (1971).

63. B. Vollmert, in "Polymer Chemistry," Springer-Verlag, New York, 1973.

64. A. H. Nissan and S. S. Sternstein, Tappi, 47, 1 (1964).

65. G. G. Allan, G. D. Crosby, J. -H. Lee, M. L. Miller and W. M. Reif, in New Bonding Systems for Paper, Proceedings of the Symposium on Man-Made Polymers in Papermaking, Helsinki, June 1972, p. 85.

66. L. S. Leiskin and R. J. Myers, in "Encyclopedia of Polymer Science and Technology," (1964), p. 246.

67. G. G. Allan, M. L. Miller, and W. M. Reif, Text. Res. J., 42, 675 (1972).

68. E. E. Treiber and D. Ehrensgard, Tappi, 57, 69 (1974).

69. M. L. Miller, in "Ionic Bonding in Rayon Nonwovens, Ph. D. Thesis, University of Washington, 1972.

70. Y. Fahmy and H. El-Saied, Holzforschung, 28, 61 (1974).

71. H. A. Krassig, Tappi, 46, 654 (1963).

72. G. G. Allan and A. N. Neogi, J. Adhesion, 3, 13 (1971).

73. Union Carbide Corporation, Tech. Brochure, Silanes, 1968.

74. K. Ward, Jr., in "Chemical Modification of Papermaking Fibers," Dekker, New York, 1973.

75. G. G. Allan and T. Mattila, Tappi, 53, 1458 (1970).

76. W. B. Achneal and A. A. Vaidya, J. Soc. Dyers Colour., 85, 404 (1969).

77. A. A. Vaidya, Cellul. Chem. Technol., 5, 417 (1971).

78. K. V. Sarkanen, in "Lignins" (K. V. Sarkanen and C. H. Ludwig, eds.), Interscience, New York, 1971, pp. 95-155.

79. C. H. Ludwig, in "Lignins" (K. V. Sarkanen and C. H. Ludwig, eds.), Interscience, New York, 1971, pp. 327-344.

80. G. G. Allan, P. Mauranen, A. N. Neogi, and C. E. Peet., Tappi, 54, 206 (1971).

81. G. G. Allan, P. Mauranen, and A. N. Neogi, Pap. Puu, 53, 371 (1971).

82. J. Stofko, E. Zavarin, and A. P. Schniewind, Division of Cellulose, Wood, and Fiber Chemistry, 167th National Meeting, American Chemical Society, Los Angeles, Cal., March 1974.

83. J. E. Laine, K. Akagane, G. G. Allan, and E. Passot, Division of Cellulose, Wood, and Fiber Chemistry, 167th National Meeting, American Chemical Society, Los Angeles, Cal., March 1974.

84. G. G. Allan, G. Bullock, and A. N. Neogi, J. Polym. Sci., 11, 1759 (1973).

85. G. G. Allan, K. Akagane, and J. E. Laine, unpublished results, 1974.

86. G. G. Allan, P. Mauranen, A. N. Neogi, and C. E. Peet, Chem. Ind. (London), 623 (1969).

AUTHOR INDEX

This is a Cumulative Index: Part I, pages 1 to 416; Part II, pages 417 to 632. Numbers in parentheses are reference numbers and indicate that an author's work is referred to although his or her name is not cited in the text. Underlined numbers give the page on which the complete reference is listed.

A

Abbott, G. M., 567(5), 568(5), 571(9), 575

Abbott, N. J., 84(63), 105

Abere, J. F., 512(53), 522

Abrams, E., 208(50), 210(50), 223

Achhammer, B. G., 368, 386

Achneal, W. B., 622(76), 631

Adam, N. K., 420, 444, 482, 493

Adamson, A. W., 144(169), 161, 420(5), 426(5), 444, 600(35), 630

Adderley, A., 82(56), 87, 105, 106

Adelman, R. L., 578(5), 597(5), 629

Adler, A., 489, 494

Ahlbrecht, A. H., 514(64, 65), 522

Airy, J. B., 73(28), 104

Akagane, K., 605(50, 51), 613(59), 623(83), 626(85), 630, 631

Alexander, P., 115(29), 134(137), 157, 160

Alfrey, G. F., 252(55), 284(55), 289

Alger, K. W., 366(15, 16), 376(16), 385

Algera, L., 115(25), 119, 157

Algie, J. E., 154(217), 163

Ali, M. A., 140(149), 161

Allan, A. J. C., 428, 445

Allan, A. J. G., 3(13), 32(13), 46(13), 47(13), 63

Allan, F. J., 611(58), 631

Allan, G. G., 578(1, 2, 5, 8, 9), 580(2, 17, 18), 582(18), 584, 585(18), 592, 597(5), 605(50, 51), 610(57), 611(58), 614(62), 615(65, 67), 616(67), 622(72, 75, 86), 623(80, 81, 83), 625, 626(85), 629, 630, 631, 632

Alter, H., 147(196), 149(196), 162

Alurkar, R. H., 151(203a), 163

Ambady, G. K., 70(6), 104

Amberg, C. H., 539(32), 560

American Association of Textile Chemists and Colorists, 252(62), 253(77), 255(62), 258(97), 266(97), 290, 291

American Enka Company, 584(22), 629

American Society for Testing and Materials, 98(116, 117, 118), 107, 195, 205(37), 222, 223, 256(82), 258(82), 290

Amick, C. A., 154(224), 163

Amin, S. A., 71, 104

Amis, M. M., 195(9), 222

Amonton, G., 2, 3, 63, 86, 106

Anderson, C. A., 148(190), 162

Anderson, D. B., 69(3), 74, 75(3), 78(3), 104

SUBJECT INDEX

This is a Cumulative Index: Part I, pages 1 to 416; Part II, pages 417 to 632.

Printed and bound by CPI Group (UK) Ltd, Croydon, CR0 4YY

23/10/2024

01778224-0009